Minoru Ozima

Geohistory
Global Evolution of the Earth

English by Judy Wakabayashi

Springer-Verlag
Berlin Heidelberg New York
London Paris Tokyo

Professor Dr. MINORU OZIMA
Geophysical Institute
University of Tokyo
Tokyo 113, Japan

With 40 Figures

ISBN 3-540-16595-9 Springer-Verlag Berlin Heidelberg New York
ISBN 0-387-16595-9 Springer-Verlag New York Berlin Heidelberg

Library of Congress Cataloging-in-Publication Data. Ozima, Minoru. Geohistory : global evolution of the Earth. Translated from Japanese. Includes index. 1. Earth – Origin. 2. Isotope geology. I. Title. QB631.036 1987 551.7 86-33903

This work is subject to copyright. All rights are reserved, whether the whole or part of the material is concerned, specifically those of translation, reprinting, re-use of illustrations, broadcasting, reproduction by photocopying machine or similar means, and storage in data banks. Under § 54 of the German Copyright Law where copies are made for other than private use, a fee is payable to "Verwertungsgesellschaft Wort", Munich.

© Springer-Verlag Berlin Heidelberg 1987
Printed in Germany

The use of registered names, trademarks, etc. in this publication does not imply, even in the absence of a specific statement, that such names are exempt from the relevant protective laws and regulations and therefore free for general use.

Media conversion, printing, and bookbinding: Konrad Triltsch, Würzburg
2132/3130-543210

Preface

One of the most striking developments in earth science in the twentieth century is the use of radiogenic isotopes as a time marker to record geological phenomena. Besides enabling radiometric age determination, which is perhaps their most common use, radiogenic isotopes provide unparalleled information on the evolution of the earth and meteorites. They are a prime source of information in understanding the origin of the solar system. In the late 1970's, yet another powerful approach has emerged in the form of computers with an extremely large memory capacity. It is hardly an exaggregation to say that our present understanding of the origin and evolution of the terrestrial planets is derived essentially from these two earth science disciplines. Only a few decades ago, there were no experimental approaches for testing theories on the origin of the solar system, which were themselves still "geofiction" or enigmas. Now we are in a stage where the theories are precise enough to be tested by experiments. This book describes the evolution of the earth, revealing the central role played by radiogenic isotopes in the exploration of the evolution of the earth. The book also emphasizes a recent development in theoretical studies.

The title of this book, *Geohistory*, was chosen to emphasize the two most outstanding aspects of earth science – the cardinal role of the enormous time span of planetary phenomena, and the aspect of "history". As the last chapter of the book emphasizes, studies on the evolution of the earth are analogous to studies of ordinary history in that they offer lessons for the future. As one such example, readers will see how studies of the Oklo natural reactor, which underwent a uranium fission chain reaction two billion years ago, provide an incomparable lesson in our struggle with the formidable problem of nuclear waste, which is now undermining our future.

I owe much to Dr. Raymond Jeanloz, who suggested that I write this book and also read the whole manuscript very thoughtfully and gave numerous invaluable comments. I am also highly indebted to Drs. A. Hofmann and M. Schidlowski, who reviewed the manuscript very carefully and made many helpful comments.

I benefited greatly from discussions with Dr. Kiyoshi Nakazawa on the theoretical aspects of the evolution of the earth and the solar system. Ms. Judy Featherstone Wakabayashi, my partner, prepared the English text from my original Japanese text. If the book is found readable, this is largely due to her excellent bilingual abilities and her affection for earth science.

Tokyo, Spring 1987   M. OZIMA

Contents

1 Geohistory as a Discipline 1

2 The Earth as a Planet in the Solar System 5

   2.1 Pre-Solar History 6
   2.2 Condensation Theory – From Nebular Gas to
       Crystal Particles 21
   2.3 Moon, Meteorites, and Other Planets – The Key
       to an Understanding of Early Geohistory 28

3 Evolution of the Earth 55

   3.1 The Driving Force Behind the Earth's Evolution . 55
   3.2 Composition of the Earth – The Meteorite Analogy 58
   3.3 The Layered Structure of the Earth 62
   3.4 Formation of the Layered Structure 64
   3.5 The Time of Core Formation, Based on Pb Isotopic
       Ratio Data 68
   3.6 Mantle Differentiation 71
   3.7 The Age of the Mantle 80
   3.8 Origin and Evolution of the Atmosphere and Oceans 84
   3.9 The Primordial Mantle and the Differentiated
       Mantle 93

4 Changes in the Earth's Crust 99

   4.1 Rock Magnetism and Paleomagnetism 99
   4.2 Ocean Floor Spreading, Continental Drift, and Plate
       Tectonics 110
   4.3 Exchange of Material Between the Mantle and
       the Earth's Crust 119
   4.4 Geochronology 122

5 Man and Geohistory 133

   5.1 Bolide Impacts: Mass Extinction of Life? 134
   5.2 The Fate of Radioactive Waste – The Oklo
       Phenomenon 142
   5.3 Epilog 151

Bibliography 157

Subject Index 161

1 Geohistory as a Discipline

Earth science is aimed at understanding the structure of the earth (including in a broad sense the moon, meteorites and planets) and the various phenomena that have occurred or are occurring there. All terrestrial phenomena are governed basically by the laws of physics, and in a large sense fall within the framework of physics. In certain respects, however, earth science and astronomy differ somewhat from physics in the normal sense of the term. This difference derives from the fact that in many cases terrestrial phenomena occur over an extremely long period which far exceeds man's experience. Take continental drift for example: the relative drift that has occurred between the American continents and the African and European continents took place over a period of more than 100 million years. This movement is so slow that it is impossible for man to be aware of it within his brief lifespan of 100 years at the most.

Earth science is also characterized by the fact that certain terrestrial phenomena occur only once in time. In many cases we cannot expect the phenomenon to recur in nature, much less attempt to reproduce it in the laboratory. Consider, for instance, the origin of the atmosphere and oceans (Chapter 3.8). It is believed that the atmosphere and oceans were formed as the result of degassing from within the earth in the early stages of the earth's accretion. Once this gas has escaped, it would be futile to expect another release of gas on the same scale and by the same process. The impossibility of reproducing these phenomena signifies that the earth is a unique planet among the countless stars in this universe. Ultimately, of course, it is the laws of physics which govern terrestrial phenomena. At the same time we must also recognize that some of the phenomena that are the subject of earth science studies occur only on this earth and only once in time.

A million years ago the earth must have been quite different from the earth we know today. Three or four billion years ago it must have been so different that it might be more appropriate to regard it as having been virtually another planet. In order to understand the earth, therefore, it is not enough just to investigate the earth as it is nowadays – we must comprehend it as it was from its birth right through up to the present. Let us give a specific example: through the application of the spherical harmonic analysis first worked out by Friedrich Gauss in the early 19th century, it is known that the earth's magnetic field can be represented by a magnetic dipole located more or less in the center of the earth. Based on the attempt by J. Larmor to explain the origin of the magnetic field in the sun, in the 1950's E. Bullard suggested that the earth's dipolar magnetic field is the result of the movement of the liquid inside the earth's core. Numerous subsequent studies have evolved a more detailed theory, known as the dynamo theory, on the origin of the earth's magnetic field. This alone, however, is not sufficient for a full understanding of the earth's magnetic field. Recent paleomagnetic research (Chapter 4) has revealed that the earth's magnetic field is by no means fixed and constant, but that it has constantly reversed its north and south polarity in roughly equal cycles. This reversal is an intrinsic property of the earth's magnetic field, and any theory on its origin which is unable to provide a full explanation of this reversal is inadequate.

As this example demonstrates, an understanding of earth science, that is, research into earth science, is insufficient if it is merely an understanding of the present state of the earth. True understanding is not achieved until the earth is viewed as having changed constantly ever since its birth in the solar system. However, it is impossible to observe phenomena extending over billions of years, so earth science research calls for special methods. Tracing the record of these phenomena over billions of years is the primary goal. Since the 19th century attempts have been made to elucidate geological phenomena by using fossils, and much important information has been gathered. In recent years fossil observations have become even more detailed through the introduction of the electron microscope, and new areas are being opened up in the study of microfossils in deep ocean sediment. Though paleontological methods are playing a major role in throwing light on the evolution of the earth, the use of biological fossils is inevitably limited. Paleontological methods are virtually useless for clarifying the history of the earth in the Precambrian period (the period before the Cambrian period began approximately 670 million years ago), for which almost no biological fossils exist. The Precambrian period accounts for roughly six-sevenths of the 4500 million years of the earth's history, and is the period which witnessed the formation of the basic structure of the earth, such as the core, atmosphere, and oceans. In order to investigate this period, for which there are no biological fossils, we must obtain some kind of fossil to take the place of biological fossils. In the 1950's an entirely new type of "fossil" was developed, using long-life radioactive elements such as ^{238}U and ^{40}K and their daughter elements produced through radioactive decay. In Chapter 4 we will take a look at several examples of elucidating geohistory by using these radioactive elements. This is the field of geochronology. Geochronological methods using radioactive elements form the nucleus in geohistorical research. Radioactive decay elements produce daughter elements at a constant rate in strict accordance with the law of radioactive decay, and so they are "fossils" which have an extremely precise time marker that acts as a built-in clock.

Along with radiogenic isotope fossils, paleomagnetism is another powerful method for studying the history of the earth. The past geomagnetic field is preserved in the form of remanent magnetism as if it had been stamped on volcanic and sedimentary rocks. Comparing the directions of this fossil magnetism, i.e., the direction of the remanent magnetism, reveals the relative movement of blocks of land. We will look at these paleomagnetic methods in Chapter 4. With the amazing progress made in large computers in recent years, a completely new method known as numerical experiments is making its appearance. By using the latest large-memory computers, to a certain extent it has become possible to trace phenomena that occurred over a long period far beyond the empirical time sphere of man, as well as global phenomena far beyond the scale of the laboratory. Numerical experiments using computers have a virtual monopoly on such issues as the evolution of the solar system and the origin of planets. A detailed discussion of numerical experiments lies beyond the scope of this book, but we will survey some results of these experiments in Chapter 2.

Here we have looked at geohistory as a discipline aimed at throwing light on the true image of the earth. As well as being the subject of such purely scientific research, geohistory also has a very direct connection with us human beings in its application. As a discipline which investigates the "history" of the earth, in this sense geohistory shares a common aspect with the study of history. Let us try to explain this through an example. It is said that people read a lot of history in times of upheaval. Apprehension about the future and expectations and uneasiness in the face of the unknown lead man to seek lessons in the past. In the latter half of the twentieth century the remarkable progress in industry and technology has given rise to a great number of environmental problems. The accelerated use of fossil fuels is causing the amount of carbon dioxide in the atmosphere to rise more or less linearly. The increased carbon dioxide in the air has the effect of heating the atmosphere as the result of infrared ray absorption, and in the near future this may lead to a catastrophic rise in atmospheric temperature. However, it is extremely difficult to make a prediction on this matter. Probably the most scientific choice currently available is to seek a similar phenomenon in the earth as it was in the past, and to predict the future based on this. The final chapter in this book will consider the relation between geohistory and the human race.

2 The Earth as a Planet in the Solar System

The earth was born in the solar system about 4500 million years ago. Since then it has undergone such changes as the formation of the core and atmosphere and mantle-crust differentiation, finally evolving into its present state. It would be no exaggeration to say that the intrinsic direction of this evolution had already taken shape when the earth broke away from the solar nebula to become a planet. For instance, suppose that in line with conventional thought the formation of the earth was completed after the dissipation of the solar nebula. In this case heat would have been easily released from the earth into space as infrared radiation. It is predicted that under these conditions the earth would have had a relatively low temperature. Suppose on the other hand that the formation of the earth reached completion before the dissipation of the solar nebula, as a Kyoto group of scientists headed by C. Hayashi has maintained recently. This would mean that the gravitational force of the earth would attract a thick primary atmosphere from the solar nebula. It is predicted that this primary atmosphere would have prevented infrared radiation from the earth, and that the earth would have reached quite a high temperature in its early stages. An understanding of the environment that produced the earth, i.e., an understanding of the early solar system, is essential to understand the evolution of the earth. As another illustration of this, it would be impossible on this earth, which has experienced over 4500 million years of constant evolution, to find rocks that record its primeval state. Thus it is also impossible to seek any clue to the age of the earth – the most fundamental quantity in earth science – in the rocks found on earth at present. Our current conclusions about the age of the earth are based in essence on the dating of meteorites, some of which are believed to preserve intact the state of the early solar system. In this chapter we will center our discussion on meteorites, which are the key to the evolution of the earth and which provide the most important clue to the state of the early solar system.

2.1 Pre-Solar History

Currently the solar system contains as many as 83 different elements. H and He account for roughly 99.9% of the total number of atoms in the solar system (Table 2.1). Practically all of the elements in the solar system are concentrated on the sun, and represent more than 99.9% of the mass of the solar system. Viewed from the elemental composition of the solar system as a whole, the earth consists mainly of oxygen and nonvolatile elements (e.g., Fe, Mg, and Si, which do not readily become gases), with the nonvolatile elements accounting for less than 0.1% of the elements constituting the solar system. Whereas discussions of the sun and the solar system virtually ignore all elements except for H atoms, the interaction of many different elements is of intrinsic importance in the origin and evolution of the earth and other planets. Most of the elements constituting the solar system and the earth were created before the formation of the solar system, but a few elements have been formed since then through the decay of radioactive elements. For instance, it has been concluded that nearly all (more than 99%) Ar, which accounts for about one percent of the earth's atmosphere, became a component of the atmosphere as the result of the ^{40}K within the earth decaying to form ^{40}Ar (^{40}K $\rightarrow$ ^{40}Ar) after the formation of the earth, and that this then degassed from within the earth. Apart from such radiogenic elements, all other elements already existed before the formation of the solar system. The question is how and when were they formed.

a) Origin of Elements

The origin of elements goes back to the big bang, which is thought to have occurred 10 000–20 000 million years ago when the Universe was created. The origin of the elements as the result of the big bang was first advocated by G. Gamow in 1948. Gamow took the view that in the early stages of cosmic creation the Universe was in a state of extremely high temperature and high pressure, and that protons, neutrons, electrons, and neutrinos were in a state of equilibrium. When the Universe began to expand, the temperature fell and this state of equilibrium was upset. Some of the neutrons underwent β-decay (nuclear disintegration that releases negative electrons), while protons captured neutrons and formed deuterons. The repetition of this process of β-decay and neutron-capture led to the successive formation of heavy elements. According to Gamow, it took only about 20 minutes to create all of the elements in current existence, truly justifying the name "big bang". Later in 1965 Aro Penzias and Robert Wilson observed microwaves impinging isotropically to the earth from all directions in space, thus providing empirical support for the big bang theory. The wavelength of the microwaves observed by Penzias and Wilson was equivalent to 3 K of black body radiation. They interpreted this as meaning that owing to the Doppler effect the wavelength of the light radiated at the time of the big bang increased as the Universe expanded, until finally it became equivalent to the current black body radiation of 3 K. This 3 K of radiated light is the afterglow of the big bang.

However, the big bang does not fully explain the origin of elements. Even if many kinds of elements formed from neutrinos, and electrons as well as protons and neutrons were created by the big bang, most of these elements would have disintegrated again in the big bang, and would not have remained in significant amounts. Hence it is now thought that elements with an atomic number of 6 or more were created through nuclear reactions within the stars from light elements formed in the big bang.

Hydrogen and the other light elements formed in the big bang were scattered about the Universe and finally came to form stars. Under the effect of their own gravity the stars began to contract gradually, thus leading to a rise in temperature. When the temperature in the center of each star exceeded several million degrees the hydrogen atoms finally combined and formed He atoms – that is, nuclear fusion occurred. Further increases in pressure and temperature inside the stars caused nuclear fusion reactions in which He atoms combined and reactions in which C atoms combined, and this process was repeated to gradually produce heavy atomic nuclei.

However, these nuclear fusion reactions can proceed no further than the formation of Fe nuclei at the most. Unilateral fusion of atomic nuclei will eventually destabilize the nuclei in terms of energy. This is similar to the collapse that occurs if too many blocks are piled up until they become unstable. Even if the Fe nuclei cause further nuclear fusions and produce heavy nuclei, these would be unstable and soon collapse. At this point those that are formed through nuclear fusion and those that become too heavy and collapse balance each other and form a state of equilibrium, so that nucleosynthesis will proceed no further. Fe atomic nuclei can be regarded as the end of the fusion reactions. In actual fact, the cosmic abundances of elements (Fig. 2.1) decrease linearly (on the logarithmic scale) as the atomic number decreases, but suddenly take an upward turn at the Fe nuclei, where they reach the maximum value.

Merely piling up atomic nuclei (fusion reaction) is not sufficient to create atomic nuclei heavier than Fe nuclei. Neutrons must lend a hand here. Neutrons are produced within stars in the process of the He fusion reactions. Since neutrons have no electric charge there is no electrical repulsion, and they are incorporated within the atomic nucleus compa-

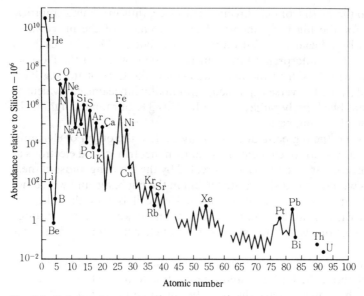

Fig. 2.1. Relative elemental abundance (Si ≡ 10^6) in the solar system. Th and U are shown for the values at 4.5 billion years ago. Data are from Ross and Aller (1974)

ratively easily. This is the phenomenon known as neutron capture reaction. In this manner atomic nuclei capture relatively low-energy neutrons and simultaneously release electrons (β-decay), thus gradually growing into heavy atomic nuclei. This process is regarded as a relatively slow reaction, and is known as the s-process (slow process).

However, the neutron capture reaction does not continue ad infinitum. If the atomic nucleus is loaded with too many neutrons, it will again become unstable and collapse. Bi is regarded as the limit of the neutron capture reaction, i.e., the s-process nucleosynthesis. In order to create an atomic nucleus heavier than Bi, it is necessary to irradiate it with an extremely powerful neutron beam and to drive in the neutrons before the atomic nucleus breaks. This kind of intense neutron beam is thought to materialize when a star explodes, i.e., at the time of the explosion of a supernova. Owing to the extremely rapid progress of the neutron capture reaction at this time, it is known as the r-process (rapid process).

The majority of the elements existing in the Universe are thought to have been created through nuclear fusion reactions and the neutron capture reactions known as the r- and s-processes. However, it is difficult to create such atomic nuclei as ^{112}Sn through these reactions, and it is thought that they were probably created by a reaction in which protons were captured directly. This is known as the p-process. It is presumed that a few light elements (D, Li, Be, B) were created neither in the big bang nor in an s- or r-process, but by interstellar gas being irradiated by

Pre-Solar History

cosmic rays, or that they were created within the atmosphere of the early sun.

Summing up the above, the elements that compose the solar system and our earth at present were nearly all formed through nuclear reactions inside stars, with the exception of H (thought to have existed from the beginning of the Universe), the He that was formed from H in the big bang, and a small number of light elements (D, Li, Be, B). As will be discussed in a later section, countless numbers of stars have been involved in these nuclear reactions, forming elements over a period of more than 10 000 million years. From astronomical observations it is known that heavy elements are relatively more abundant in comparatively new stars than in old stars. This is because there were few heavy elements in the universe when the old stars formed, and this observed fact is regarded as support for the nucleosynthesis described above. Moreover, it has been ascertained empirically that elements having a half-life of less than several tens of millions of years, such as ^{129}I and ^{244}Pu, once existed in meteorites and the earth. Radioactive elements become virtually extinct over several half-lives. Consequently, the nucleosynthesis that produced these elements must have occurred immediately before the formation of the meteorites and the earth (a time equivalent at the most to several times the half-life of these elements). From the isotopic composition (^{238}U/^{235}U = 137.8; current value) of atomic nuclei with a relatively long half-life, such as U (half-life of ^{238}U: $T_{1/2} = 4.47 \times 10^9$ years, half-life of ^{235}U: $T_{1/2} = 7.04 \times 10^8$ years), it is possible to estimate the length and time of the nucleosynthesis that formed thse atomic nuclei. Based on these results, it is inferred that U and similar elements with a long half-life were formed more than 1000 million years ago over a period of several hundred million years. Detailed calculations will be shown in the following section. This is one of the major results of the new field known as cosmochronology.

b) Cosmochronology

Section (a) discussed briefly the age of elements. Here we will proceed with a more quantitative discussion of the subject. As will be stated later, generally speaking the isotopic compositions of elements composed of solar material are extremely uniform. The isotopic compositions of elements, such as ^{238}U/^{235}U, for example, are ordained by the nucleosynthetic processes that created these elements. Conversely, by using the isotopic composition of elements such as U or Th, it is possible to infer the nucleosynthesis that created the element. This is an oft-used device in earth science, with the cause being sought from the result, and is generally known as the inverse method. Let us commence from a very general discussion.

Focus now on the element N. Let us call the i-th isotope of this element N_i. N_i can be a radioactive isotope that generally undergoes radioactive decay. Consider the case when N_i is formed through nucleosynthesis. The temporal change in N_i at this time is

$$\frac{d}{dt} N_i(t) = P_i \cdot p(t) - \lambda_i \cdot N_i(t). \tag{2.1}$$

Here

λ_i: the decay constant of isotope N_i
P_i: the proportion of N_i which is synthesized in the nucleosynthesis
$p(t)$: the intensity of the nucleosynthesis.

The first term on the right-hand side shows the amount of N_i that is synthesized in unit time, and the second term shows the amount of N_i that decreases further through radioactive decay. $p(t)$ shows the extent to which each synthesis occurs, and can be regarded as describing the degree of the stellar explosion that resulted in the nucleosynthetic reaction. In general the manner in which elements are formed differs depending on the manner of explosion of the star. Therefore the isotope N_i is not necessarily formed in the same proportion P_i in the explosion of all stars. However, this hypothesis can be viewed as reasonable as regards isotopes that are formed through the same kind of nucleosynthesis, e.g., the r-process or the s-process. The time axis is shown in Fig. 2.2. In this figure $p(t)$ is shown schematically on the vertical axis.

Formally integrating Eq. (2.1), the result is

$$N_i(t_0) = P_i \cdot \exp(-\lambda_i t_0) \int_0^T \exp(\lambda_i \tau) \cdot p(\tau) \cdot d\tau. \tag{2.2}$$

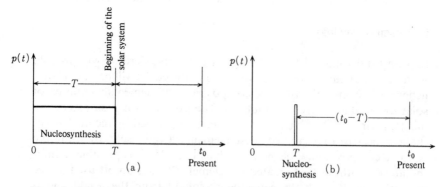

Fig. 2.2. Schematic representation of nucleosynthesis models. **a** a continuous nucleosynthesis model; constant nucleosynthesis from $t = 0$ to $t = T$. **b** a sudden nucleosynthesis model; nucleosynthesis at $t = T$

Pre-Solar History

In Eq. (2.2), the p(t) on the right-hand side and the upper limit T of the integration, i.e., the duration T of the nucleosynthesis, are the unknown quantities that we are seeking. The N_i on the left-hand side is the total abundance of the isotope N_i in the solar system at present (t_0), and even though it can be observed in principle, it is not practically possible to estimate its value. Hence, we have no means to link Eq. (2.2) to the observed value N_i and thus to find p(t) and T. Let us refer here to the usual practice in the physical approach – this is the method of transforming the experimental value and the observed value into ratios and comparing them with the theoretical equation. Toward this end, let us formulate an equation that is identical to Eq. (2.2) for another isotope N_j of the element N.

$$N_j(t_0) = P_j \cdot \exp(-\lambda_j t_0) \int_0^T \exp(-\lambda_j \tau) \cdot p(\tau) d\tau. \qquad (2.3)$$

Dividing (2.3) by (2.2):

$$\frac{N_j(t_0)}{N_i(t_0)} = \frac{P_j}{P_i} \exp\{-(\lambda_j - \lambda_i) \cdot t_0\} \times \frac{\int_0^T \exp(\lambda_j \tau) \cdot p(\tau) d\tau}{\int_0^T \exp(\lambda_i \tau) \cdot p(\tau) d\tau}. \qquad (2.4)$$

As will be discussed in the next section, with a few exceptions the isotopic compositions of the elements composing the material found in the solar system are uniform on the whole. Thus even a tiny fragment of solar material, such as a meteorite or an earth rock, can be regarded as representing the whole of the solar system as far as its isotopic composition is concerned. Hence the left-hand side of Eq. (2.4) is a measurable quantity. It is also possible to estimate P_j/P_i on the right-hand side semi-empirically. This alone, however, is not enough to enable us to find p(t) and T from Eq. (2.4). Some assumptions must be made and a model set up in order to do this. Next we will consider specific examples of this in two extreme cases.

First let us suppose that the nucleosynthesis has proceeded uniformly throughout the period T (Fig. 2.2a). This is known as the continuous nucleosynthesis model. In Eq. (2.1) this is equivalent to letting

$$\begin{aligned} p(t) &= \text{const.} & 0 \leq t \leq T \\ p(t) &= 0 & t > T. \end{aligned} \qquad (2.5)$$

A more concrete picture of this can be conjured up by imagining that the frequency of the stellar explosions that created the elements was uniform throughout the period T. The value of (N_i/N_j) at the time $(t = T)$ when the nucleosynthesis came to an end can then be calculated from Eqs. (2.4) and (2.5):

$$\left(\frac{N_j}{N_i}\right)_T = \left(\frac{P_j}{P_i}\right) \cdot \exp\{-(\lambda_j - \lambda_i) \cdot T\} \frac{\frac{1}{\lambda_j}(e^{\lambda_j T} - 1)}{\frac{1}{\lambda_i}(e^{\lambda_i T} - 1)}. \quad (2.6)$$

As the other extreme case, suppose that the nucleosynthesis occurred virtually instantaneously at a certain time (T) (Fig. 2.2b). Here p(t) can be written as follows:

$$p(t) = A \cdot \delta(t - T).$$

Here δ is the delta function and A is the constant corresponding to the intensity of the reaction. Since the comparison with the observed values uses Eq. (2.4), however, the actual figure for A is not necessary. The value at the time t can be calculated from Eq. (2.4):

$$\left(\frac{N_j}{N_i}\right)_t = \frac{P_j}{P_i} \cdot \frac{\exp\{-\lambda_j(t - T)\}}{\exp\{-\lambda_i(t - T)\}}. \quad (2.7)$$

Here we have hypothesized two extreme nucleosynthesis models and deduced the general equations corresponding to each of these. Using these equations, let us now find T in the case of ^{238}U and ^{235}U.

Continuous Nucleosynthesis Model

In Eq. (2.6), let $N_i = {}^{238}$U and $N_j = {}^{235}$U.

$$\frac{N_j}{N_i} = ({}^{235}U/{}^{238}U)_T = \frac{(P_{235}/\lambda_{235})}{(P_{238}/\lambda_{238})} \frac{1 - \exp(-\lambda_{235}T)}{1 - \exp(-\lambda_{238}T)}. \quad (2.8)$$

Here λ_{235} and λ_{238} are the decay constants of ^{235}U and ^{238}U respectively. T is the time at which the nucleosynthesis finished. As mentioned in section (a), the fact that radioactive decay elements with an extremely short half-life, such as ^{129}I and ^{244}Pu, were included in solar material indicates that the end of the nucleosynthesis that created the solar material occurred almost simultaneously to the formation of the solar system. Thus T can be regarded as 4550 million years ago – the age of the solar system. Since currently $({}^{235}U/{}^{238}U)_P = 1/137.8$ (the subscript P indicates the present), 4550 million years ago – that is, at the time when the nucleosynthesis finished –

$$({}^{235}U/{}^{238}U)_T = ({}^{235}U/{}^{238}U)_P \times \frac{\exp(4.55 \times 10^9 \times \lambda_{235})}{\exp(4.55 \times 10^9 \times \lambda_{238})} = 0.3165.$$

Pre-Solar History

Estimated on a semi-theoretical basis, the value of P is

$P_{235}/P_{238} = 1.42$.

Substituting these values in Eq. (2.8) produces the following equation:

$$0.3165 = 1.42 \times \frac{1.55125 \times 10^{-10}}{9.8484 \times 10^{-10}}$$
$$\times \frac{1 - \exp(-9.8485 \times 10^{-10} \times T)}{1 - \exp(-1.55125 \times 10^{-10} \times T)}$$

$T \cong 8 \times 10^9$ (years).

This means that the r-process nucleosynthesis that produced U lasted for 8000 million years between the beginning of the solar system (4 550 million years ago) and a time 8000 + 4550 = 12 550 million years ago. Again using the isotopic ratio data for U, let us next consider the "sudden synthesis model" that supposes that the nucleosynthesis was a single "sudden" reaction.

Sudden Nucleosynthesis Model

In Eq. (2.7) let us take the present time (t_0) for t. Consequently ($t_0 - T$) indicates the time from when the "sudden" nucleosynthesis occurred up until the present time. When the values corresponding to ^{238}U and ^{235}U are inserted for P_i, P_j, i, and j respectively and the current value of 137.8 for the ratio $^{238}U/^{235}U$ is inserted, we obtain

$t_0 - T = 6.4 \times 10^9$ years.

According to the sudden nucleosynthesis model, therefore, the nucleosynthesis would have occurred approximately 6400 million years ago.
As can be seen in the above example, the time at which the nucleosynthesis is thought to have occurred can almost double depending on how the model is formulated. Both of these are strictly models and are mere approximations, thus resulting in this discrepancy. However, since these two models correspond, so to speak, to the two extremes of the actual nucleosynthesis, the time of the nucleosynthesis can be regarded as lying somewhere between the solutions in these two extreme cases.
Tracing in further detail the circumstances in which the nucleosynthesis took place would mean hypothesizing a more detailed nucleosynthesis model, but here we are faced with the necessity of introducing more unknown parameters (in the sudden synthesis model the only unknown parameter was T) that constrain the model. Hence, more conditions are needed to solve the problem. In addition to the $^{235}U - ^{238}U$ pair described in the example above, currently more com-

plex models using such pairs as ^{238}U–^{232}Th or ^{187}Re–^{187}Os as their condition have been proposed.

c) Composition of the Solar Nebula

The elements created inside countless stars over more than 10 000 million years have been scattered constantly throughout space. These elements formed nebulae, one of which became our solar nebula. Opinion is currently divided on how the sun and planets were born from the solar nebula. We will leave a detailed discussion of this to textbooks (Hayashi et al. 1985) on the theory of the formation of the solar system, and focus our discussion here on points that are directly related to the earth.

Why did the solar nebula begin to break away and separate from the nebulae? The ^{26}Al that formerly existed (by now it has completely disintegrated into ^{26}Mg and no longer exists) in meteorites provides a hint to the answer to this question. The half-life of ^{26}Al is only 700 000 years. Thus the fact that ^{26}Al existed in meteorites indicates that the nucleosynthesis occurred at most several times the half-life of ^{26}Al prior to the formation of meteorites, i.e., the formation of the solar system. If the formation of ^{26}Al was the result of the explosion of a supernova, it is thought that the resultant shock caused the nebular gas to commence gravitational contraction, making the solar nebula break away and separate. Very recently, however, it has been pointed out that ^{26}Al may have been formed not only through the explosion of a super nova, but also through the explosion of a nova, and so it is no longer possible to conclude that the birth of the solar nebula was necessarily the result of the explosion of a supernova.

Nebular gas is formed by the elements that have been created within a countless number of stars being scattered about, and the elemental composition of nebular gas is therefore thought to be well-homogenized. In particular we can expect that the isotopic compositions, which are hardly affected at all by any physical or chemical differentiation, will be highly uniform. In fact it is known that apart from a few exceptions all of the observable material within the solar system (earth material, meteorites, moon rocks, the surface of the sun, etc.) has more or less uniform isotopic compositions (excluding radiogenic nuclei which were added after the formation of the solar system). The uniformity of the isotopic compositions of solar material has been ascertained through isotopic analyses of meteorites and earth rocks, which have been carried out with great zest since 1950, and through spectroscopic studies of the sun's corona. This uniformity has provided conclusive proof that the sun, earth, meteorites, and other bodies were formed from one and the same solar nebula. This discovery is so vital that it can be regarded as the starting point of modern earth science and planetology.

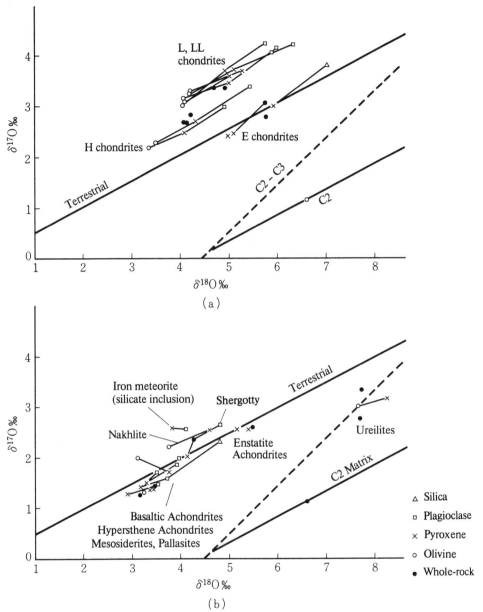

Fig. 2.3. Oxygen isotope variation in various meteorites. **a** chondrites, **b** achondrites and stony-iron meteorites. $\delta^{17}O \equiv [(^{17}O/^{16}O)_{sample}/(^{17}O/^{16}O)_{SMOW} - 1] \times 1000$ and similar for $\delta^{18}O$. SMOW means Standard Mean Ocean Water. Note that C2–C3 chondrites (anhydrous minerals) lie on a line with a slope of unity (shown by a *dotted line*), showing that their oxygen isotopes have different origin from those of other meteorites or the earth. (After Clayton et al., 1976)

The elements in solar material have virtually identical isotopic compositions, but there are a few exceptions, most typical of which is O. With the cooperation of T. Mayeda, R. Clayton measured the oxygen isotopic ratios of several meteorites. The results were far different from those anticipated, and are shown in Fig. 2.3.

As physical and chemical changes occur, $^{18}O/^{16}O$ and $^{17}O/^{16}O$ undergo mass fractionation. Let us consider, for example, the process whereby H_2O evaporates. Since lighter isotopes evaporate more easily, as evaporation proceeds the heavier isotope ^{18}O is gradually left behind in the water, while the amount of the lighter isotope ^{16}O in the vapor increases. This is known as mass fractionation. The greater the difference in mass, the more marked is this effect. The mass fractionation between ^{18}O and ^{16}O can be shown mathematically as being twice that between ^{17}O and ^{16}O, mirroring the difference in mass. Thus, when plotted as in Fig. 2.3, the resulting mass fractionation is shown by a straight line with a slope of 1/2, reflecting the fact that the mass fractionation of $^{18}O/^{16}O$ (the horizontal axis) is twice that of the mass fractionation of $^{17}O/^{16}O$ (vertical axis). Mass fractionation is caused not only by such physical processes as the evaporation described above, but also by chemical reactions. Hence solar material that originally possessed identical $^{18}O/^{16}O$ and $^{17}O/^{16}O$ isotopic ratios produces mass fractionation by undergoing countless subsequent physical and chemical processes, and it is anticipated that the values will be scattered along the straight line with a slope of 1/2 shown in Fig. 2.3. In reality, all of the many objects of earth material that have been measured so far do line up neatly along a straight line with a slope of 1/2 (shown in Fig. 2.3 by the solid line). As stated in the previous section, if meteorites were formed from the same highly homogeneous solar nebula as the earth, at the time of their birth they would have possessed the same oxygen isotopic composition $^{18}O/^{16}O$ as the earth. Even if they underwent some kind of physical or chemical process after their birth, it is expected that they would still line up on the same straight line with a slope of 1/2 that of the earth.

However, the results found by Clayton and Mayeda ran completely contrary to these expectations. As shown in Fig. 2.3, the results for several carbonaceous chondrites are scattered along a straight line whose gradient is almost 1. Clearly this shows that the oxygen constituting these chondrites is not merely the product of mass fractionation from the oxygen that is a component of the earth. If the difference between the two is not due to mere physical or chemical processes, we must seek the cause in the difference between the nucleosyntheses that produced them.

Subsequent research has revealed that meteorites of different types all have their own individual oxygen isotopic ratio (Fig. 2.3). This is interpreted as meaning that the oxygen constituting solar material consists of a combination of oxygen that has at least two different origins (i.e., is formed by differing nucleosyntheses) and oxygen that was produced from this through various different mass fractionation effects (all of these can be connected to each other by a line with a gradient of 1/2). At this stage, however, it is not known whether there are only two kinds of this different type of oxygen or whether oxygen having other origins also exists.

The research into oxygen isotopes by Clayton and Mayeda has provided convincing proof that solar material is not always necessarily "extremely homogeneous". The "heterogeneity" of solar material had already been proposed by D.C. Black and R.O. Pepin in 1969 based on the isotopic composition of rare gases in meteorites. In several carbonaceous chondrites Black and Pepin detected an independent Ne component in which ^{22}Ne alone is extremely condensed, though in very tiny amounts, and they labeled this component Ne–E. Naturally, Ne–E has an isotopic composition that differs markedly from that of the Ne in the earth's atmosphere and the normal Ne found in meteorites.

As discussed in the previous section, the solar nebula is thought to be well-mixed, homogenized nebular gas that contracted owing to self-gravity and broke away to become separate. On the other hand, the isotopic compositions of oxygen and Ne and a few other elements show that the material within the solar system has differing isotopic compositions, i.e., these elements are of a different origin. How should we view this contradiction? Before answering this question, we must add a note of explanation about isotopic anomalies.

As already stated, the elements constituting solar material have highly uniform isotopic compositions, but the isotopic compositions of a few materials, including oxygen and Ne, display a disparity that cannot be explained merely by mass fractionation. Such disparities are known as isotopic anomalies. Currently isotopic anomalies have been detected in Mg, Ca, Ti, V, Kr, Sr, Xe, Ba, Nd, and Sm in addition to O and Ne. What is important here, however, is the fact that these isotopic anomalies have been detected only in certain minerals in meteorites. Ne–E, which led to the discovery of isotopic anomalies, is found in the carbonaceous material of carbonaceous chondrites and in spinel ($MgAl_2O_4$) and apatite [$Ca_3(PO_4)_2CaF_2$]. Clayton and Mayeda found an isotopic anomaly of O in the spinel and pyroxene in the Allende meteorite, while no isotopic anomalies were detected in any other minerals in the same meteorite. Judging from the results of research so far, it seems difficult to link the isotopic anomaly of a specific element to any specific mineral phase.

Without doubt the isotopic anomalies that have been found since the discovery of Ne–E indicate strongly that the solar nebula was by no means homogeneous. Taking an overall view, however, these isotopic anomalies can be regarded precisely as that – anomalies. It seems feasible to conclude that the composition of solar material and the early solar nebula that created this material was approximately homogeneous. The isotopic anomalies include such short-lived isotopes as ^{26}Mg and ^{129}I. We can interpret this as meaning that at least for these isotopes the nucleosynthesis that was the cause of the isotopic anomalies occurred immediately before the formation of the solar system. The same applies for other isotopes also. It is highly likely that they were created as the result of the explosion of a supernova immediately before the formation of the solar system, and that they mixed locally in the primitive solar system material, thus resulting in anomalies. Another possible interpretation is that material which has an isotopic anomaly was formed a long time before the formation of the solar nebula. It is also possible to take the view that these "pre-solar" materials existed in the nebular gas as solid particles, so that they have preserved the isotopic compositions that existed long before the formation of the solar system, without being homogenized with the nebular gas.

We can conclude that solar material – the sun, planets, and meteorites – created from the solar nebula that is thought to have been more or less homogenized as a first approximation has identical isotopic compositions. On the other hand, it is expected that the elemental compositions will differ as the result of fractionation, mirroring the physical and chemical conditions in which each element is placed. As far as the elemental composition of the earth as a whole is concerned, there is a marked scarcity of rare gases and volatile elements such as H, N, and S. The same applies to meteorites. On the other hand, the abundance of nonvolatile elements (although there is no continuous and qualitative borderline distinguishing them from volatile elements), which do not readily become gases even at high temperatures, differs little amongst different materials.

These are compared in Fig. 2.4, with the elemental composition observed within the sun's corona plotted on the vertical axis and the elemental composition of meteorites (carbonaceous chondrites) plotted on the horizontal axis. An excellent correlation is evident between the two. The sun accounts for more than 99.9% of the mass of this system, and it is thought that the sun's corona is mixed with the inner layers quite well as the result of convection, so there is no objection to regarding the elemental composition observed in the corona as approximating that of the whole solar system. The similarity between the elemental composition (relative) of meteorites and the elemental composition of the sun's corona – limited of course to nonvolatile elements – indicates the possibility of inferring the elemental composition of the whole solar system from precise measurements of the elemental composition of meteorites. There are limits to the accuracy with which the elemental composition of the sun's corona can be identified by optical methods. Laboratory

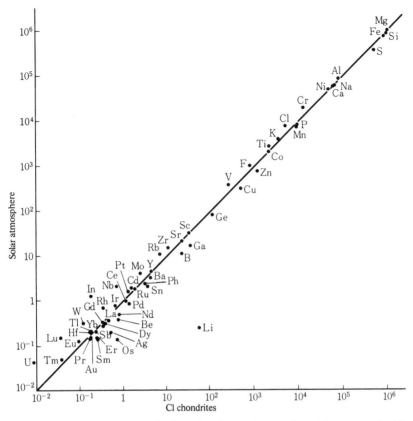

Fig. 2.4. Elemental abundance in the solar photosphere (*ordinate*) is compared with that in C1 chondrites (*abscissa*) (Si ≡ 10^{-6}). Except for highly volatile elements, there is an excellent correlation between them. Data are from Anders and Ebihara (1982) for C1 chondrites and from Ross and Aller (1974) for the sun

analyses of the elemental composition of meteorites guarantee greater accuracy. E. Anders and his colleagues carried out high-quality chemical analyses of a number of meteorites, and attempted to estimate the solar abundances of elements on the basis of these results. Naturally a random analysis of meteorites would be useless. Certain kinds of meteorite have clearly undergone metamorphism as the result of weathering on the earth. There are also some, such as irons, which have a special element composition vastly different from that of the solar system as a whole. Taking these points into consideration, Anders and his colleagues chose the C1 chondrite as being most typical of the elemental composition of the whole solar system. The solar abundances shown in Table 2.1 are the values calculated by Anders and his colleagues from the analyzed values

Table 2.1. Elemental abundances in the solar system. (Anders and Ebihara 1982)

Element	Abundance	Element	Abundance	Element	Abundance	Element	Abundance
$_1$H	2.72×10^{10}	Ti	2400	Ru	1.86	Dy	0.398
He	2.18×10^9	V	595	Rh	0.344	Ho	0.0875
Li	59.7	Cr	1.34×10^4	Pd	1.39	Er	0.253
Be	0.78	Mn	9510	Ag	0.529	Tm	0.0386
B	24	Fe	9.00×10^5	Cd	1.59	$_{70}$Yb	0.243
C	1.21×10^7	Co	2250	In	0.184	Lu	0.0369
N	2.48×10^6	Ni	4.93×10^4	$_{50}$Sn	3.82	Hf	0.176
O	2.01×10^7	Cu	514	Sb	0.352	Ta	0.0226
F	843	$_{30}$Zn	1260	Te	4.91	W	0.137
$_{10}$Ne	3.76×10^6	Ga	37.8	I	0.90	Re	0.0507
Na	5.70×10^4	Ge	118	Xe	4.35	Os	0.717
Mg	1.075×10^6	As	6.79	Cs	0.372	Ir	0.660
Al	8.49×10^4	Se	62.1	Ba	4.36	Pt	1.37
Si	$\equiv 10^6$	Br	11.8	La	0.448	Au	0.186
P	1.04×10^4	Kr	45.3	Ce	1.16	$_{80}$Hg	0.52
S	5.15×10^5	Rb	7.09	Pr	0.174	Tl	0.184
Cl	5240	Sr	23.8	$_{60}$Nd	0.836	Pb	3.15
Ar	1.04×10^5	Y	4.64	Sm	0.261	Bi	0.144
K	3770	$_{40}$Zr	10.7	Eu	0.0972	Th	0.0335
$_{20}$Ca	6.11×10^4	Nb	0.71	Gd	0.331	U	0.0090
Sc	33.8	Mo	2.52	Tb	0.0589		

of C1 chondrites. The reason for selecting the C1 chondrite from among the many types of meteorite was that it is regarded as being the most undifferentiated and primitive meteorite in terms of its mineralogical and chemical composition. For example, as well as high-temperature minerals such as olivine, C1 chondrites contain a lot of volatile matter such as water and gases, showing that they have not been affected by thermal metamorphism and other disturbances after their formation.

If the earth was formed from a homogenized solar nebula just as the meteorites, sun, and planets were, we can expect that the average elemental abundances on earth will resemble those of the whole solar system. In fact, our current knowledge about the elemental composition of the earth has been gained primarily from analogies with meteorites. This is the meteorite analogy approach. We will leave a detailed discussion of the validity of the meteorite analogy and of the elemental composition (estimated values) of the earth to Chapter 3, and in the next section we will deal with the question of the process by which the earth was formed from the solar nebula.

2.2 Condensation Theory – From Nebular Gas to Crystal Particles

a) Molecules in the Primitive Solar Nebula

It is assumed that part of the primitive solar nebula that had just broken away and separated as the result of gravitational contraction reached quite a high temperature (several thousand K). At such a high temperature atoms were in a dispersed state, and had not formed chemical molecules. Eventually radiation from the surface of the primitive solar nebula led to a loss of energy and the temperature began to fall. Finally it dropped to a level at which a certain kind of molecule could exist, and molecules formed within the solar nebula. The application of classic thermodynamics gives a good idea of the manner in which molecules form when the temperature of a high-temperature gas falls. Put in qualitative terms, molecules with a high melting point are the first to condense and form a solid phase, followed at lower temperatures by molecules with a lower melting point. This theory is known as the condensation theory.

In order to deal with this theory more quantitatively, it is necessary to know (i) the composition of the gas, and (ii) the equilibrium constant (as a function of the temperature, of course) of the chemical reaction that forms each molecule. For (i) we can assume the solar abundances, and for (ii) we can use values measured in the laboratory. This work was carried out systematically by A.S. Grossman and J.W. Larimer. In order to perform this calculation, Grossman considered 300 molecules ($3H_2 + N_2 = 2NH_3$, $2H_2 + O_2 = 2H_2O$ etc.), and assumed that condensation occurred with each of their chemical reactions being maintained in a state of equilibrium throughout. He showed that as the high-temperature primitive solar nebula cooled, firstly corundum (Al_2O_3), then perovskite ($CaTiO_3$) and other high-temperature minerals solidified, and then such Ca and Al compounds as spinel ($MgAl_2O_4$) and diopside ($CaMgSi_2O_6$) solidified. A further drop in temperature led to the appearance of plagioclase [(CaA, NaSi) $AlSi_2O_8$] and magnetite (Fe_3O_4), and finally such solid or liquefied molecules as ice and ammonia or methane were formed (Fig. 2.5).

Let us now develop our argument in a little more detail in line with Grossman's method. The solar abundances consist overwhelmingly of hydrogen atoms, as is evident from theoretical studies of nucleosynthesis and from observations of interstellar clouds. Consequently, as a first approximation, we can take the view that the number of atoms in the solar nebula was approximately equal to the number of hydrogen atoms. If the total number of atoms is N and the number of H atoms is N_H, then

$$N \cong N_H$$

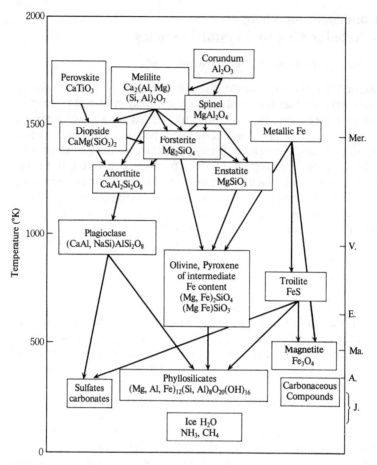

Fig. 2.5. Condensation sequence of elements in the early solar nebula with falling nebula temperature. *Ordinate (left)* indicates temperature at which minerals were formed in the solar nebula. In the *right hand ordinate*, each terrestrial planet is assigned to the respective temperature where the planet accreted. (After J. Wood, 1979)

Similarly, the number Nx of atom x can be written as

$$Nx = \frac{A(x)}{A(H)} \cdot N_H. \tag{2.9}$$

Here A(x) and A(H) are the relative abundances of X atoms and H atoms in the solar nebula, i.e., the solar abundance (also known as the cosmic abundance, see Fig. 2.1 for details).

While the temperature of the primitive solar nebula was still extremely high, all elements existed as dispersed simple atoms, but as the temperature fell various compounds were formed through chemical reaction. Focusing on the hydrogen atom, let us now consider conditions at this time. The total number of hydrogen atoms within the primitive solar nebula can be expressed as follows as the sum total of these many compounds.

$$N_H = N_H + 2N_{H_2} + 2N_{H_2O} + 3N_{NH_3} + 4N_{NH_4} + \ldots . \qquad (2.10)$$

In reality, only a few hydrogen compounds exist in nature in significant quantities, so we can restrict the compounds on the righthand side of Eq. (2.10) to these realistic compounds. The question of the relative proportion in which hydrogen compounds are formed within the nebula can then be determined univocally in Eq. (2.10) when given the temperature and elemental composition of the nebula (assuming that it was in a state of chemical equilibrium). In order to demonstrate this concretely, let us consider the case of NH_3 molecules.

The chemical equilibrium of N, H and NH_3 is drescribed by the equilibrium constant K as

$$N + 3H \rightleftharpoons NH_3$$
$$K = P_{NH_3}/P_N \cdot P_H^3 . \qquad (2.11)$$

Here P_x indicates the partial pressure of molecule x. Since the equilibrium constant is determined as a function of temperature, if we are given the temperature and the total number of H and N atoms that existed originally, we will be able to estimate the proportions in which N, H and NH_3 are able to exist. Equation (2.11) uses the equation of state for gases

$$P = n \cdot RT$$

(P: partial pressure, n: molecules per unit volume, T: temperature, R: gas constant)

to produce the following equation

$$N_{NH_3} = K \cdot N_N \cdot N_H^3 \cdot (RT)^3 . \qquad (2.12)$$

Similar equations to Eq. (2.11) can be written also for other molecules apart from NH_3.

Our ultimate goal is to determine the relative abundance of compounds at a given temperature T. This corresponds to solving Eq. (2.10) and finding each term on the right-hand side when given the temperature T of the nebula, with the relative abundances [Eq. (2.9)] of the primitive solar nebula as the initial condition. In addition to Eq. (2.9), Eq. (2.12) is also used as the condition for solving Eq. (2.10). As another initial condition, we shall make the theoretical conjecture that $P = 10^{-3}$ atm as the gas pressure of the primitive solar nebula. Let us now assume n number of compounds in Eq. (2.10) as realizable hydrogen compounds. Thus the combined total of these n number of compounds and N_H^0, (n + 1) is the unknowns in Eq. (2.10). By contrast, a total of (n + 1) equations can be formulated [Eq. (2.10)], and the Eqs. (2.12) can be written for n compounds. Hence if the value of the equilibrium constant K for each compound has been found empirically as a function of temperature, it will be possible to solve Eq. (2.10) completely. Grossman (1972) has carried out the above calculation by considering the 20 elements that are most abundant in the solar system and the 300 kinds of realistic compounds made from the bonding of these.

In Grossman's calculations it was assumed that all reactions took place in a state of equilibrium. This is the situation when the temperature falls at an extremely low rate. If the temperature fell quite rapidly, it would become impossible to maintain a state of equilibrium even on a local scale. In response to this, the condensation theory needs to deal also with states of nonequilibrium. Consequently, the mineral phases that solidify as the temperature falls would also come to differ slightly. Here, however, we will not go into the condensation theory in any more detail, but will confine ourselves to a few comments.

The condensation theory assumed a high-temperature primitive solar nebula as its starting point. The high temperature means that all atoms are completely dispersed and no molecules have been formed yet. Even if molecules that had been created prior to the formation of the primitive solar nebula remained, the condensation theory assumes that they would dissociate again within the high-temperature gas. At present, however, we have no real understanding of the mechanism enabling the primitive solar nebula to reach such a high temperature. As stated in the previous section, some chondrites contain at least two kinds of oxygen that are thought to be of obviously different origins. If the temperature of the primitive solar nebula rose sufficiently to evaporate all of the molecules, the oxygen atoms would have been well homogenized, and it would be too much to expect different kinds of isotopes to be preserved intact within the same meteorite. Meanwhile, minerals (mineral phases rich in Ca and Al) in some chondrites have been observed to have a layered structure in which the inner part consists of high-temperature minerals and progressively lower temperature minerals as one proceeds outwards. This is interpreted as a reflection of the systematic condensation from a high to a low temperature.

The records remaining in meteorites present both of the following aspects: (i) the primitive solar nebula formerly consisted entirely of gases, and (ii) some solid particles were mixed in heterogeneously. Some researchers view these apparently contradictory observed facts as mean-

Condensation Theory – From Nebular Gas to Crystal Particles

ing that the rise in the temperature of the primitive solar nebula was a local occurrence, and that it did not rise sufficiently in some parts, so that solid particles were also mixed in.

b) Formation of Planetesimals

In the discussion above we have assumed that the solar nebula was an independent mass of gases. At this stage we can only speculate as to why it should have separated as a mass from the nebulae that extend throughout the whole universe. The explosion of a supernova may have been what triggered this separation. Recently there have been reports of gases that can be regarded as solar nebulae having been observed beside supernova.

Setting aside the mechanism by which the gaseous nebula was formed, however, its subsequent dynamic development can be surmised if we suppose that a gaseous sphere with a mass roughly equivalent to that of the present solar system was formed. This evolutionary process can be divided into the following three cases depending on the size of the angular momentum of the gaseous sphere:

i. Angular momentum is zero: Under the effect of its own gravity the entire gaseous nebula would eventually concentrate in the center, and only an object corresponding to our sun would be formed. No planets would be formed.

ii. Gentle angular momentum: Most of the mass would concentrate in the middle of the gaseous sphere and form a "sun". Some of the gas would remain as a nebula surrounding the "sun".

iii. Larger angular momentum: The gaseous sphere would split into two "suns" with roughly the same mass. This corresponds to the case of binary stars.

It is obvious that our solar system corresponds to the case in (ii) and that further planets were formed from the gas remaining around the "sun". Let us now systematically explain the process by which the sun was formed.

As the nebular gas that formed the primitive sun revolved in the center it gradually spread out into a disc shape on the side that is perpendicular to the rotation axis (Fig. 2.6a). Gas and particles are concentrated gradually on the equatorial plane of the disc. At this time the temperature of the nebula dropped considerably, and crystallized particles and gases were mixed in the nebula (Fig. 2.6b–d). Eventually the gases and particles gradually concentrated on the equatorial plane. When the density reached a critical value (theoretically estimated as $\varrho \simeq 10^{-8}\,\text{g cm}^{-3}$ in the vicinity of the earth's orbit) the gases and particle clouds became gravitationally unstable, and fragmented into count-

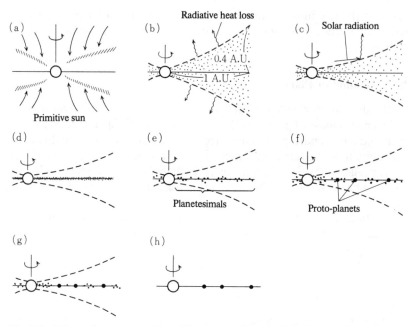

Fig. 2.6. Schematic representation of formation of terrestrial planets. **a** formation of a proto-sun in a slowly rotating nebula; **b–c** grain formation in the solar nebula and their subsequent sedimentation on the equatorial plane; **d** gravitational fragmentation of the dust layer and formation of numerous planetesimals: **e–g** accretion of proto-planets; **h** dissipation of solar nebula. (After Nakazawa, 1981)

less numbers of small gas and particle clusters typically of about 10 km in size (Fig. 2.6e). This is the phenomenon known as fragmentation of the dust layer, and was theoretically concluded independently in about 1970 by V.S. Safronov, C. Hayashi and P. Goldreich and W.R. Ward.

The masses of gases and particles that were formed as the result of this fragmentation are known as planetesimals. The planetesimals continually collided with each other as they underwent Kepler motion in the gravitational field of the sun. If the planetesimals combined completely each time they collided, they would grow with each collision and eventually develop into planets (Fig. 2.6f–h). This is the basic scenario for the formation of planets, and is common to all researchers. However, by using a figure smaller than 1 (perfect combination) as the sticking probability at the time of collision, or by using a different method of taking the initial velocity distribution of the planetesimals, an infinite number of different scenarios can be formulated, even though the end result – the formation of the planets – is the same. It goes without saying that in order to formulate a truly scientific and creative theory many more experimental and observational constraints are necessary.

Few empirical and observed data are available at present, so it is difficult to reach an unequivocal conclusion about the correctness of theories on the formation of the planets. The fact that the success of the scenario – supposing that a plausible explanation for the various phenomena can be found – does not necessarily prove its validity is the weak point of this kind of theory. In order to overcome this drawback, inductive methods, that is, efforts to seek out records of the past in nature, take on an especially vital significance in the study of the earth's evolution.

Meanwhile, let us consider cases in which the deductive approach is used to formulate a scenario. Although all eventually leading to the formation of the planets, different scenarios result in very different conclusions about the subsequent evolution and development of the planets. Let us give an example: so far nearly all theories on the formation of the planets have assumed that the planetesimals existed in a vacuum and collided with each to grow into large planets. This is the planet formation theory represented by Safronov and others. The assumption of a vacuum is based on the conclusion that, as will be discussed in Chapter 3, the earth's atmosphere is a secondary atmosphere resulting from degassing from the interior of the earth, and no primary atmosphere existed on earth as a remnant of the solar nebula. Although the earth's atmosphere is undeniably of a secondary origin, this does not necessarily completely rule out the possibility that a primary atmosphere left over from the solar nebula surrounded the earth in the very early stages of its evolution. Rather, if we consider that the planets were born in the solar nebula, it would be far more natural to take the view that the solar nebula adhered to the planets, attracted by their gravity. Currently the discussion of this issue is being considered from two standpoints, one which establishes a vacuous space as the initial condition for the planet formation scenario, and one which assumes a space still replete with the solar nebula. So far no observed facts that decisively refute either viewpoint have been discovered. However, this selection is of primary importance to our view of the evolution of the earth. If the earth grew in a vacuum, it would have continuously lost heat through radiation from its surface, and the temperature would not rise. On the other hand, if it grew within the nebula the surrounding nebular gas would have acted as a cushion to prevent radiation, and would result in a marked increase in temperature, as has been pointed out recently by H. Mizuno and other scientists. In the latter case it is presumed that a considerable part, particularly the upper part, of the earth in its early stages was in a molten state. Whether to adopt the "accretion in a vacuum" viewpoint or to view the earth as having formed within gases is a vital choice governing our understanding of the evolution of the earth. Seeking observed facts to form a basis for this selection is the central task in the study of the earth's evolution, i.e., in geohistory. If one adopts the viewpoint that the earth accreted within the nebula, one is forced to conclude that the earth was in a high-temperature, molten (upper part) state in its early stages. At present we do not possess any reliable observed facts showing that the primeval earth was in a molten state on a large scale. On the other hand, it is known that on the surface of the moon there is a large magma ocean that was formed in the very early stages (approximately 4500 million years ago) of the moon's evolution. Was the earth in a molten state in its initial stages? This vital question is of primary importance to geohistory, and is also linked to the fundamentals of solar system and planet formation theory.

As is clear from this example, the selection of the earth formation scenario plays a decisive role also in our picture of the subsequent development of the earth and its current state. Let us consider another example: as will be discussed in detail in a later chapter, the earth at present has a distinct layered structure consisting of crust, mantle, and core. If we assume that the planets were formed in a comparatively short time from a solar nebula with a uniform composition, the newly formed earth would also have had a more or less uniform composition, and so it would be natural to conclude that the formation of the layered structure occurred after the formation of the earth. On the other hand, if planet formation occurred over an extremely long period, it is conceivable that the components of the solar nebula would have differentiated, and that considerable component changes would also have occurred in planets in the early-formed central part and in the later-formed outer part. In this case, it would mean that the layered structure of the earth was formed during the process of the formation of the earth. Merely from a theoretical scenario of the earth's formation it is not possible to judge whether the layered structure of the earth is a primary or a secondary one. We must seek clues in our observations of the present earth. This, too, is a central problem in geohistory, and will be discussed in detail in Chapter 4.

2.3 Moon, Meteorites, and Other Planets – The Key to an Understanding of Early Geohistory

The earth is a planet with a radius of approximately 6400 km. By contrast, the radius of the moon is less than 2000 km, and the parent bodies of most meteorites, even large ones, are thought to be less than 500 km in radius. (The meteorites found on earth are fragments that fell to earth by chance when their parent bodies collided and were destroyed). These differences in size have a decisive effect on the subsequent evolution and development of each body. The driving force behind the evolution of the earth and planets is the energy released through the nuclear disintegration of radioactive elements contained in tiny amounts in the material composing the earth and planets, as well as the primordial heat due to accretion and core formation. U (^{235}U, ^{238}U), ^{232}Th and ^{40}K are the main sources of the former energy. These are all long-lived nuclides with a half-life of thousands of millions of years. In Table 3.1 we will show the energy released by these nuclides in the unit time.

These nuclides contained in meteorites, the moon and other planets, and the parent bodies of meteorites slowly release energy, which is then stored as heat. The size of the body is the deciding factor in how the heat is stored. Because of the very low heat conductivity of rocks on bodies as large as the earth, the heat can escape only very slowly, and such bodies remain hot for very long periods of time. The heat stored in the earth's interior eventually becomes the source of energy for igneous and volcanic activity or for orogenic movements. On the other hand, heat escapes easily from the surface of smaller objects, such as the parent bodies of meteorites, and is not stored internally. Owing to the radioactive energy stored inside, the earth and other planets undergo repeated orogenic movements, and are constantly evolving and developing. By contrast, the small parent bodies of some meteorites do not have sufficient energy stored to cause igneous or volcanic activity, and so from the moment of their birth they exist as "dead bodies". We can regard the parent bodies of such meteorites (e.g., carbonaceous chondrites), and consequently the meteorites that were broken off from them and fell down to earth, as preserving intact the conditions at the time of their birth. Hence meteorites are a valuable material that could be described as fossils of the early solar system for earth scientists. To a lesser degree, the moon, which is far smaller than the earth, is in a situation similar to that of the parent bodies of meteorites. H. Urey took the view that as a "dead planet" the moon preserved intact the initial state of the earth as a "dead planet", and strongly advocated that studies of the moon should be promoted in order to gain an understanding of the earth itself. Such claims were a major force behind the subsequent moon exploration program and the Apollo program. As the following example illustrates, as well as being one of the most vital areas of research in planetology, the study of the moon and meteorites is also indispensable to an understanding of the earth itself, particularly the early stages of the earth's evolution. Already a vast amount of research has been carried out into the moon and meteorites, and numerous papers and books have been written. The task of describing all or even some of the results of these studies lies far beyond the scope of this book. Here we will focus on the aspect of how research into the moon and meteorites has contributed to an understanding of the evolution of the earth.

a) Meteorites – Dating the Formation of the Solar System

As described in the previous section, with a few exceptions the isotopic compositions of meteorites, the moon, and earth material can be regarded as being approximately identical. This suggests strongly that meteorites (and their parent bodies), the moon and the earth were formed from a solar nebula that had a well-mixed and uniform composition. Based on this, chemical analyses of meteorites are the most reliable method currently available for inferring the average chemical composition of the earth. Virtually the only earth material on which we can lay our hands now is crustal matter. The earth's crust was formed as the result of igneous and volcanic activity after the earth underwent material differentiation. The average composition of the earth's crust differs from that of the earth as a whole. On the other hand, it is known that such material differentiation as igneous and volcanic activity did not occur on most meteorites and their parent bodies because of their small size, and hence that they preserve the chemical composition of the original solar nebula (excluding the volatile components that could not be condensed in the parent bodies of meteorites). If the earth is viewed as a whole, it can be regarded approximately as possessing a chemical composition (for nonvolatile components) that closely resembles that of the solar nebula. We will leave to the next chapter a specific conclusion as to the average composition of the earth as concluded from the chemical analyses of meteorites. Also, in a later section we will show estimates of the age of the earth as an example illustrating the effectiveness of meteorite research in throwing light on the evolution of the earth. Below we will discuss the age of meteorites.

The "living planet" Earth has undergone repeated igneous and volcanic activity and formed crustal rocks, while metamorphism and erosive and weathering action have in turn broken down these rocks. It is impossible to find on the earth today rocks that were formed at the time of the birth of the earth, and attempts to measure the age of earth material will not result in finding the age of the earth. Estimates of the age of the earth are forced to rely on indirect methods. The most reliable of these methods is to use the meteorite-earth analogy and conclude that the age found experimentally for meteorites is the same as the age of the earth. We will leave a more detailed discussion of estimates of the age of the earth to the next section, and discuss here the dating of meteorites.

Similar to the dating of terrestrial samples, the dating of meteorites uses radiometric dating such as the K–Ar and Rb–Sr methods. A full account of radiometric dating techniques is given in Chapter 4.4. Here we will discuss the dating of meteorites, taking the Rb–Sr method as an example.

The Rb existing in nature consists of two isotopes – ^{85}Rb (27.8%) and ^{87}Rb (72.2%). Rb is a radioactive element, and ^{87}Rb decays into ^{87}Sr $+\beta^-$ with a half-life ($T_{1/2} = 4.88 \times 10^{10}$ years). Suppose that a meteorite was formed t years ago. The word "formed" is very ambiguous, but for now let us consider it to be the time when the minerals constituting the meteorite crystallized. Naturally the minerals can be considered to have also contained some Sr and Rb. In nature Sr consists of four isotopes – ^{88}Sr, ^{87}Sr, ^{86}Sr, and ^{84}Sr. All four of these are stable isotopes, and do not themselves change. However, as time passes ^{87}Sr gradually accumulates owing to the decay of the ^{87}Rb within the crystals. From the radioactive decay equation this change can immediately be written as

$$^{87}\text{Sr} = (^{87}\text{Sr})_i + {}^{87}\text{Rb}\,(e^{\lambda t} - 1). \tag{2.13}$$

Here the subscript i shows the amount of ^{87}Sr (also known as the initial Sr value) that already existed when the crystals were formed (t years ago), and the quantity with no subscript shows the current value (t = 0). λ is the decay constant of ^{87}Rb, and $\lambda = 1.42 \times 10^{-11}\,\text{y}^{-1}$. Here a word of caution is necessary concerning the general methods of isotopic earth science. When carrying out experiments, it is very difficult to measure precisely the absolute values of isotopes. Using a mass spectrometer, however, it is a comparatively easy matter to find their relative values (isotopic ratio). This applies not only to isotope measurements, but also to physical and chemical measurements in general – relative measurements can be carried out more precisely and easily than absolute measurements. Hence in isotope earth science it is normal to express not the absolute values of isotopes, but their ratios.

When both sides of Eq. (2.13) are divided by the stable isotope ^{86}Sr in order to express it using isotopic ratios, the result is

$$\frac{^{87}\text{Sr}}{^{86}\text{Sr}} = \left(\frac{^{87}\text{Sr}}{^{86}\text{Sr}}\right)_i + \frac{^{87}\text{Rb}}{^{86}\text{Sr}}(e^{\lambda t} - 1). \tag{2.14}$$

In this equation the subscript i is the value t years ago, and all other quantities are current values. ^{86}Sr is a stable isotope that does not change over time, so $^{86}\text{Sr} = (^{86}\text{Sr})_i$. In addition to the date t that we are seeking, Eq. (2.14) contains another unknown quantity, $(^{87}\text{Sr}/^{86}\text{Sr})_i$. This is called the initial ^{87}Sr/^{86}Sr isotopic ratio, and contains information of great importance to earth science, just as the age t does (Chap. 4). Since Eq. (2.14) has two unknown quantities, it is not possible to find the age t by analyzing just one sample. In order to find t, we must analyze the ^{87}Sr/^{86}Sr and ^{87}Rb/^{86}Sr ratios of at least two samples and insert these values in Eq. (2.14), and then solve the simultaneous equations to find t and $(^{87}\text{Sr}/^{86}\text{Sr})_i$.

A note of warning is necessary here. If we suppose that each mineral in the meteorite crystallized from a common melt, since the isotopes are completely identical chemically, the isotopic ratios distributed in each mineral can also be regarded as being identical to the values of the melt. This is the state known as isotopic equilibrium. If no such equilibrium exists, then different minerals will have different $(^{87}Sr/^{86}Sr)_i$ values, and it will not be possible to find the age t using Eq. (2.14). The question of whether or not an isotopic equilibrium exists, i.e., whether or not the t obtained by solving Eq. (2.14) is meaningful as the formation age of the meteorite, can be determined by applying the isochron method discussed below.

The actual method of finding the age using Eq. (2.14) is to separate at least three different minerals from the meteorite and analyze the $^{87}Sr/^{86}Sr$ and $^{87}Rb/^{86}Sr$ ratios of each mineral. These are then plotted on a graph with the $^{87}Sr/^{86}Sr$ ratios on the Y axis and the $^{87}Rb/^{86}Sr$ ratios on the X axis (cf. Fig. 2.7). As is obvious from Eq. (2.14), if each mineral is formed simultaneously (at the identical age t) and if a state of isotopic equilibrium existed at that time [the minerals had the same $(^{87}Sr/^{86}Sr)_i$ ratio], then the points of measurement for each mineral should lie on the graph (Fig. 2.7) in a straight line. The slope of this straight line would be $e^{\lambda t} - 1$ ($\cong \lambda t$, if $\lambda t \ll 1$).

The age t can be found from the slope of the straight line. All material having a common initial Sr isotopic ratio $(^{87}Sr/^{86}Sr)_i$ and formed at the same time t will lie on this straight line. This straight line is called an isochron, and Fig. 2.7 is called an isochron plot.

In the discussion above we have considered minerals as the samples. However, it is also possible to draw up an isochron plot using individual meteorites as the sample. When regarding a whole rock as one sample, the age thus determined is known as the whole rock age. By contrast, the age found from minerals is called the mineral age. Note here that the significance of the age t found from Eq. (2.14) differs depending on the type of sample used. If only the minerals constituting the meteorite are used as the sample, the age found is the age at which the minerals crystallized, and its significance is clear. In the case of ordinary chondrites (also known as equilibrium chondrites), which are meteorites that have undergone intense metamorphism and contain unmistakable evidence of the whole meteorite having recrystallized, we can take the view that the constituent minerals were formed as the result of metamorphic action. The mineral age of meteorites in this case indicates the time at which the meteorite underwent metamorphism.

If experiments are carried out on several meteorites as whole rock samples and an isochron is obtained, it is no easy matter to reach a conclusion as to what kind of meteorite-formation process does the age t found from Eq. (2.14) correspond. The isochron shows that each mete-

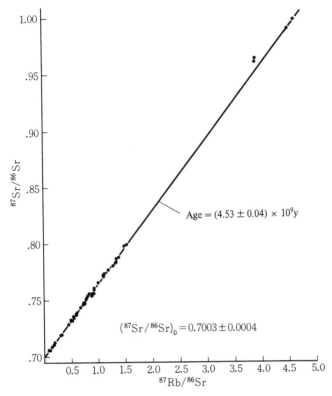

Fig. 2.7. Rb–Sr isochron plot for chondrite whole rock samples which gives an age of 4.53×10^9 yr and an initial ratio $^{87}Sr/^{86}Sr$ of 0.7003. Minerals separated from chondrites also lie on the same isochron (not shown in figure). (After Wetherill, 1975)

orite possessed the same initial Sr-isotopic ratio $(^{87}Sr/^{86}Sr)_i$ t years ago. If Rb exists in the material the $^{87}Sr/^{86}Sr$ isotopic ratio will change as the result of ^{87}Rb disintegration. How it changes will depend on the Rb/Sr ratio within the material. In cases in which solid material condenses from the gaseous solar nebula, we expect that in general each mineral component would have a different Rb/Sr ratio owing to elemental differentiation. When viewed like this, it seems to be most logical to seek the environment capable of preserving the common initial Sr-isotopic value $(^{87}Sr/^{86}Sr)_i$ in the actual gaseous solar nebula, which is homogenized and more efficiently mixed. Thus it may be appropriate to consider the whole rock age of meteorites as the time at which their constituent elements, i.e., solid crystal particles, crystallized and separated from the solar nebula.

Figure 2.7 shows the Rb–Sr isochron plot found for the whole rock samples of a number of chondrites. Regardless of type, the meteorites line up extremely well on the isochron. The isochron age of $t = 4.53 \times 10^9$ years can be found from the slope of the isochron. Furthermore, the isochron for minerals that have separated from some chondrites is virtually identical to that found for the whole rock samples. From these empirical facts it is concluded that all meteorites came into existence about 4500 million years ago as solid material that had separated from the solar nebula, and that a certain kind of meteorite (equilibrium chondrites) underwent metamorphism-recrystallization immediately after separating from the solar nebula.

In the same manner as the Rb–Sr method, it is possible to carry out isochron dating using the combination of ^{147}Sm–^{143}Nd ($T_{1/2} = 1.06 \times 10^{11}$ years). The results accord extremely well with the value found by the Rb–Sr method.

Another slightly indirect method – historically this method led to the first dating of meteorites – is to use the Pb-Pb method to conclude that the time of meteorite formation was $t = 4.55 \times 10^9$ years.

The very slight discrepancy among these three results is thought to be the result of the uncertainty of such decay constants as ^{87}Rb and ^{147}Sm, rather than to any difference in their significance to earth science. The fact that the three independent datings conform extremely well within the range of experimental error enables us to conclude that $t = 4.53 \times 10^9$ years as the formation age of meteorites. The K–Ar method and ^{40}Ar–^{39}Ar method also have been used for many meteorites, but owing to the loss of Ar there is a large discrepancy in the age values, and it is not suitable for use in dating meteorites. On the other hand, since the loss of Ar indicates thermal disturbance of the meteorite, the K–Ar age provides vital information on the thermal history of meteorites. Some chondrites have a Rb–Sr and Sm–Nd isochron age far younger than 4500 million years, but it is thought that these show the time when collisions and secondary thermal metamorphism occurred after the formation of the meteorite.

From the above discussion we can conclude that meteorites separated from the solar nebula as "solid material" approximately 4500 million years ago. From the existence of ^{129}I and ^{244}Pu we were able to conclude that the solid material separated almost immediately after the end of the nucleosynthesis. From the amount of ^{129}Xe* of ^{129}I decay origin in meteorites it is also possible to make a quantitative estimate of the duration of the separation of the solid material after the end of the nucleosynthesis: this period is known as the formation period.

As will be explained in detail in the section on the age of the earth, it has been proved that the radioactive decay elements ^{129}I and ^{244}Pu, which have a comparatively short half-life ($T_{1/2} \sim 10^7 \sim 10^8$ years), once existed in meteorites and on the earth. Such short-lived radioactive elements have virtually become extinct during the 4500 million years since the formation of the solar system, and only the stable daughter elements formed through their decay can be observed now. Such nuclides with a short life are known as extinct nuclides. As discussed below, extinct nuclides offer unique information on meteorites and the conditions surrounding the formation of the solar system. The nuclides used as the sources of this information include ^{129}I ($t_{1/2} = 1.7 \times 10^7$ years), ^{244}Pu (8.2×10^7 years) and ^{107}Pd (6.5×10^6 years). Here we will discuss the case of ^{129}I in some detail.

Let us first consider finding the time (Δt) from the completion of the nucleosynthesis until the formation of meteorites. Δt is called the formation interval of meteorites. Let us suppose that when the nucleosynthesis terminated, ^{129}I was scattered about within the solar nebula mixed in with other elements. ^{129}I gradually decreased as it disintegrated into ^{129}Xe. Finally the nebula cooled and minerals crystallized. At this time elements that form compounds, such as I, would be incorporated within the minerals as compounds, but rare gases such as Xe do not bond chemically, so would be hardly incorporated at all. As time passes, however, simultaneously with the formation of meteorite material the ^{129}I in this material (let us regard it specifically as the minerals of which the meteorite is composed) would disintegrate, and the resulting ^{129}Xe* (the asterisk indicates that it is the product of radioactive decay) would be no longer able to escape from the minerals, and would be firmly captured in this material. In general, meteorite material also contains ^{129}Xe that was directly incorporated from the nebular gas at the time of the meteorite's formation (though minute in quantity, it is the result of, for example, adsorption to crystal surfaces). As will be described in a later section ("How old is the earth?"), however, an extremely ingenious experimental method enables us to differentiate between ^{129}Xe that was formed as the result of ^{129}I disintegrating into ^{129}Xe* (asterisk denotes radiogenic origin) within the material and the ^{129}Xe that was captured when the material was formed.

Since the ^{129}Xe* observed in meteorites at present was formed as the result of the disintegration of the ^{129}I formerly incorporated within the meteorite material,

$$^{129}\text{Xe}^* = (^{129}\text{I})_0. \tag{2.15}$$

Here the subscript o shows the value at the time when the meteorite material was formed. If we use the subscript n (denoting nucleosynthesis) to express the value at the time when the nucleosynthesis finished:

$$(^{129}\text{I})_0 = (^{129}\text{I})_n \cdot e^{-\lambda_{129} \cdot \Delta t}. \tag{2.16}$$

λ_{129} is the decay constant of ^{129}I, and $\lambda_{129} = 4.1 \times 10^{-8}$ y^{-1}. Thus from Eq. (2.15) and (2.16)

$$^{129}Xe^* = (^{129}I)_n \cdot e^{-\lambda_{129} \cdot \Delta t}. \tag{2.17}$$

In order to compare the observed quantity with the equation, it is convenient to use the ratio rather than the absolute value of the observed quantity. Here we will use the stable isotope ^{127}I and rewrite Eq. (2.17) as

$$\left(\frac{^{129}Xe^*}{^{127}I}\right) = \left(\frac{^{129}I}{^{127}I}\right)_n \cdot e^{-\lambda_{129} \cdot \Delta t}. \tag{2.18}$$

Note that in Eq. (2.18) ^{129}I is a stable isotope and can be regarded as a "constant" that does not change with time. Now if it were possible to find the value of $(^{129}I/^{127}I)_n$ immediately after the nucleosynthesis, Δt could be found from Eq. (2.18), since in this equation the value on the left-hand side is the current amount contained within meteorites and is a measurable quantity. In actual practice, however, it is difficult to find the value of $(^{129}I/^{127}I)_n$. In order to find this value it is essential to know exactly how the nucleosynthesis occurred, and this is impossible at present. Therefore we will abandon the attempt to find the absolute value of Δt, and instead concentrate on finding its relative value.

Let us now consider two meteorite samples. We will formulate Eq. (2.18) for each of these meteorites and consider the difference between them. Letting $\Delta = \Delta t_1 - \Delta t_2$,

$$\Delta = \Delta t_1 - \Delta t_2 = \frac{1}{\lambda_{129}} \cdot \ln\left[\left(\frac{^{129}Xe}{^{129}I}\right)_1 \bigg/ \left(\frac{^{129}Xe}{^{127}I}\right)_2\right]. \tag{2.19}$$

Thus we cannot find the absolute value of Δt, but it is perfectly possible to find its relative value from empirical data alone. The samples used here include the Bjurbole chondrite as the standard sample, and the difference between the formation intervals relative to this meteorite is found. The results are shown in Fig. 2.8. The relative age value (unit: million years) is plotted on the horizontal axis and the Bjurbole chondrite is placed at the origin of the axes. The positive values indicate meteorites formed after the Bjurbole chondrite, and the negative values indicate ones formed earlier. As the figure reveals, nearly all meteorites measured so far were formed in an extremely short period (compared to the age of the solar system) of less than 30 million years. Emphasizing this "quick" formation of meteorites, Hohenberg et al. (1967) called it sharp isochronism. The number of meteorites that have been analyzed since then has risen considerably, but this basic trend remains unchanged.

In the discussion above the age at which meteorites formed corresponds to the time when the $^{129}Xe^*$ produced by the decay of the ^{129}I

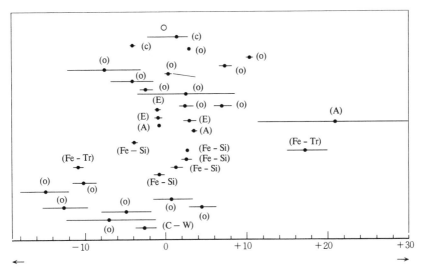

Fig. 2.8. I−Xe formation age of meteorites relative to Bjurböle chondrite (shown by *an open circle*). Ages are in millions of years, where plus and negative indicate early and late in formation relative to Bjurböle. Note that the whole span is about 30 Ma. (Compiled by N. Takaoka, 1982). C: carbonaceous chondrites; O: ordinary chondrites; E: enstatite chondrites; A: aubrites; Fe−S: Silicate inclusions in iron meteorites; Fe−Tr: troilite in iron meteorites; C−W: white inclusions in carbonaceous chondrite (Allende)

within this material has all been captured within the material and can no longer escape outside. We have understood this broadly as meaning the time of formation of the mineral crystals. Strictly speaking, however, it is by no means clear what size the minerals must reach before we can conclude that the ^{129}Xe* cannot escape. Nor is it entirely obvious how the formation interval of mineral crystals can be related to the formation interval of meteorites. Moreover, once a meteorite undergoes thermal metamorphism and the ^{129}Xe* that has been accumulated up until then is completely released and ^{129}Xe* has again begun to accumulate, the formation interval for these meteorites will correspond to the time of the metamorphism. In fact, many researchers adopt this interpretation. However, no clear correlation has been recognized between the "I−Xe formation intervals" and the type of meteorite (shows the degree of metamorphism). However, it is not entirely impossible to take the view that the differing ^{129}Xe*/^{127}I ratios of meteorites simply reflect the isotopic heterogeneity of the $(^{129}I/^{127}I)_n$ ratio. The I−Xe formation interval is a source of information of primary importance with regard to the formation of the primary solar system and meteorites, but the interpretation of this interval is not yet final.

b) The Moon

When the results of experiments on rock samples from the moon were announced, the first major surprise was the fact that their age values cover a wide range between 4500 and about 3500 million years. H. Urey had forecast that the moon froze conditions at the time of the creation of the earth, but dating shows clearly that there was a period of volcanic activity on the surface of the moon lasting for more than 1000 million years. However, no final conclusion has been reached as to whether this volcanic activity was the result of the thermal energy of radioactive decay accumulated in the moon's interior, as is the case on earth, or whether it was due to an external cause as the result of meteorite showers – most of the craters on the moon are thought to be the result of the impact of meteorites.

In moon research what is of particular interest to the study of the earth's evolution is the existence of meteorite craters, which are thought to have been formed by meteorite showers. By dating the rocks in meteorite craters, it is estimated that approximately 4000 million years ago there was a violent shower of meteorites on the surface of the moon. Because of the size and number of their meteorites, this period is known as the era of lunar cataclysm. The moon is only 380 000 km away from the earth, so that viewed from the vast solar system as a whole the two planets could almost be regarded as being located in the same position. It is only natural to think that if many meteorites fell on the moon, then even more must have fallen on the earth. As far as we know at present, however, no proof remains that would allow us to conclude that the earth also was bombarded by a violent meteorite shower about 4000 million years ago. Should we understand this as meaning that the earth was subjected to a meteorite shower just as was the moon, but that subsequent crustal activity completely erased all trace of this collision? Or should we interpret it as meaning that since the relative positions of the moon and earth approximately 4000 million years ago were completely different from their current positions, only the moon collided with the meteorite shower? At this stage no clues are available to the answer to this problem.

One interesting find recently as far as the relationship between the moon and earth is concerned is the discovery of meteorites of lunar origin. Starting with the discovery in 1969 by the 10th Japan Antarctic Research Expedition of nine meteorites in a snow field close to the Yamato base, more than 6000 meteorites have been found all over Antarctica. The meteorites that have been discovered so far cover nearly every kind of meteorite, including irons, chondrites, and achondrites. It has been pointed out that the chemical composition of the meteorite (ALHA81005) collected by the American team in 1981 very closely resembles that of the lunar breccia collected by Apollo 16. It has also been shown that the oxygen isotopic ratio of this meteorite is identical to that of moon samples, suggesting that it is highly likely to be of lunar origin. On the other hand, differences from the lunar samples that have been collected so far have been pointed out as far as its chemical composition is concerned, and if ALHA81005 really is of lunar origin it must have come from an area from which no samples have been collected so far. Since then two meteorites that are thought to be of lunar origin have been discovered in Antarctica. Even more recently, there have been reports that some meteorites classified as shergottite came from Mars, judging from their rare gas isotopic and elemental compositions as well as nitrogen isotopic composition that are very similar to those of the Martian atmosphere.

Dynamically speaking, it is very likely that objects that have broken loose from the surface of the moon will reach the earth. All that is necessary is for the energy that releases the object from the surface of the moon to exceed the speed at which the object can escape the gravitational field of the moon (escape velocity: 2.4 km/s^{-1} and 11.2 km/s^{-1} for the moon and earth, respectively). A very gentle meteorite collision would be more than enough to provide this energy. Interestingly, there are no signs on ALHA81005 of it having received a strong shock.

c) Terrestrial Planets

The planets in the solar system include terrestrial planets that are similar to the earth in their form and origin. These are Mercury, Venus, (Earth), and Mars. These planets were formed from the solar nebula just as were the sun and other planets in the solar system. Even though they were formed from a virtually homogeneous solar nebula, however, each planet has distinct characteristics. For instance, Mercury is considerably denser than the other terrestrial planets. This reflects the differences in the chemical composition of the planets. It is thought that their distance from the sun is the decisive factor behind these differences. A temperature and pressure distribution for the solar nebula immediately prior to the birth of the planets has been inferred from theoretical model calculations. Gas is more compressed the closer it is to the sun, and so the pressure is high and the temperature rises. Omitting detailed calculations, as a first approximation we will suppose that the temperature and pressure distribution of the solar nebula is described by the adiabatic compression of the gases. This situation is shown in Fig. 2.9. Whereas the temperature reaches almost 1500 K near the orbit of Mercury, which is the closest to the sun, it is less than 500 K near the orbit of Mars. It is expected that under the high temperature near the orbit of Mercury

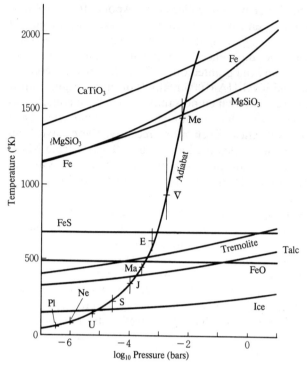

Fig. 2.9. Equilibrium condensation sequence for a system of solar elemental composition. Also shown is the adiabat in the solar nebula, on which each planet is indicated by a respective symbol, corresponding to the respective site where the planet formed. (After Lewis and Prinn, 1984)

elements cannot condense and crystallize, except for Ca, Al silicates, and Fe and Ni, which have a high melting point. This fits in well with the observed fact that Mercury has a greater density than the other terrestrial planets. J.S. Lewis attempted to explain the characteristics of the planets from this point of view.

As explained already in the previous section, as the temperature of the solar nebula falls, the minerals that crystallize change from high-temperature minerals to low-temperature minerals, followed by hydrous minerals and ice. Lewis superimposed the P–T distribution curve for the solar nebula shown in Fig. 2.9 on the condensation phase diagram (Fig. 2.5) in an attempt to explain the characteristics of the planets. As Fig. 2.9 shows, the only minerals capable of condensation in the orbit of Mercury are Fe and Ni and such high-temperature minerals as $MgAl_2O_4$, $Ca_2Al_2SiO_7$ and $CaTiO_3$. On the other hand, in the vicinity of the earth minerals with a low melting point, such as $NaAlSi_3O_8$ (albite) and $KAlSi_3O_8$ (orthoclase), are also capable of condensation.

This model by Lewis is extremely simplified, and the actual formation of the planets is bound to have been more complex. For instance, it is too simplistic to consider that the T–P distribution within the primitive solar nebula was the result of adiabatic compression alone, as hypothesized in the Lewis model. Moreover, the occurrence of the equilibrium condensation shown in Fig. 2.5 (not nonequilibrium condensation) is highly controversial. Despite these flaws, however, the Lewis model offers quite a good approximation of the fundamental characteristics of the actual planets (for example, the question of why Mercury has a greater density than the Earth and Venus).

At this stage our direct observations of the planets are limited to the atmosphere and surface of Venus and Mars. In our study of the origin and evolution of the planets, research into the structure and composition of their atmospheres is at least as important or even more important than the recovery of solid planetary material. Since the atmosphere is a fluid, it is far more well-mixed and homogeneous than solid material. Thus by taking just a tiny sample of the atmosphere and examining its composition – particularly the isotopic composition – it is possible to give an approximate estimate of the value of the whole atmosphere. The situation is fundamentally different from that of solid samples, as it is utterly impossible to estimate the average composition of a solid planet merely from an analysis of such samples, even if they weighed millions of tons. In the case of planetary probes using spacecraft, where only extremely limited observations are permitted, these circumstances have a particularly vital significance.

Table 2.2 shows the chemical compositions of some elements in the atmospheres of Venus, Earth, and Mars. It is obvious from this table that, when compared with the solar abundances (Table 2.1), both Mars and Venus resemble the Earth in the extreme scarcity of rare gases compared to such volatile components as C and N. This demonstrates conclusively that the origin of the planetary atmospheres cannot be found in the remnants of the solar nebula. As was already mentioned in Section 2.1 and as will be discussed in detail in Chapter 3 on the origin of the earth's atmosphere, the atmospheres of Venus and Mars were formed by degassing from the interior of the planets, just as the Earth's atmosphere was formed. This shows clearly that their atmospheres have a secondary origin. This signifies that the interior of these planets had a thermal history that caused the degassing. In the case of the Earth, the process of this degassing can be discussed quite quantitatively by placing certain constraints on the K-content and Ar isotopic ratio of the solid earth (Chap. 3). If it becomes possible in the future to estimate these compositions in Mars and Venus also, we can expect quite quantitative discussions of the degassing histories of these planets too.

Table 2.2. Chemical composition of planetary atmospheres

	Main component		Trace component		Isotopic composition
Venus	CO_2	90 atm. (98.1%)	CO	40 ppm	$D/H < 1/70 \sim 1/80$
	N_2	1.6 atm. (1.8±0.4%)	HCl	1 ppm	$^{40}Ar/^{36}Ar \sim 1$
	Ar	18 mb (200 ppm)	HF	1 ppb	
	H_2O	Above cloud 1000 ppm	O_2	< 1 ppm	
		Below cloud 100 ~ 1 ppm			
Earth	N_2	0.78084 atm.	CO_2	33.00 ppm	$D/H \sim 10^{-4}$
	O_2	0.20946 atm.	Ne	18.18 ppm	$^{40}Ar/^{36}Ar \sim 295.5$
	^{40}Ar	0.0934 atm.	He	5.24 ppm	$^{12}C/^{13}C \sim 89$
	H_2O	Ocean ~ 300 atm.			$^{15}N/^{14}N \sim 270$
	CO_2	Carbonatite ~ 50 atm.			$^{129}Xe/^{132}Xe \sim 1$
Mars	CO_2	5.5 mb (95.32%)	O_2	0.3%	$^{15}N/^{14}N \sim 459$
	N_2	0.16 mb (2.7%)	CO	0.08%	$^{40}Ar/^{36}Ar \sim 3000$
	^{40}Ar	0.10 mb (1.6%)	H_2O	0 ~ 80 μm (water)	$^{129}Xe/^{132}Xe \sim 2.5$

Additional trace components:
- Venus: O_3 yes
- Earth: H_2O 0 ~ 2%; Kr 1.14 ppm; Xe 0.087 ppm
- Mars: O_3 1 ppm

d) Comparative Planetology

Studying the other planets in our solar system also deepens our understanding of the Earth. The reverse situation also applies. This has led to a new approach to planetology research, known as comparative planetology. One example is the attempt to apply plate tectonics, which has been the guiding principle in solid earth science in the latter third of this century, in order to gain an understanding of the geological structure of other planets. So far research has been limited to the surface of the planets as observed by spacecraft and astronomical telescopes, and we do not know whether or not the plate tectonics theory can be applied to planets other than the Earth. The crustal movements and structures that play the lead role in plate tectonics differ from phenomena involving the whole planet, such as the layered structure – the division into core and mantle – and this is an issue in which the "individuality" of each planet is intricately involved, so the conclusions gained about the Earth may not apply to other planets.

Observations of the magnetic fields of planets place important constraints on the layered structure of the planets, and particularly on the existence of a core. This is a good example of the comparative planetology approach. At its surface the Earth has a magnetic field of about half a gauss (5×10^{-5} Tesla). The geomagnetic field can be likened to the situation if a magnetic dipole were placed in the center of the earth parallel to the earth's rotation axis. This hypothetical magnetic dipole is known as the geomagnetic dipole. The size of the magnetic dipole corresponds to a magnetic moment of $8 \times 10^{22}\,A\,m^2$. However, no such strong dipole magnetic field can be observed on Venus, which is roughly the same size as the earth. Even if a magnetic field does exist on Venus, its magnetic moment would be less than $5 \times 10^{18}\,A\,m^2$. Similarly, no clear magnetic field can be observed for Mars. On Mercury, however, which is regarded as quite different from other terrestrial planets as far as its average density and, naturally, its chemical composition are concerned, a considerable magnetic field has been observed. These differences in the magnetic fields reflect the different origin and evolutionary process of each planet. Conversely, then, it is possible to place constraints on the origin and evolution of each planet from the viewpoint of the origin of its magnetic field. The magnetic fields of planets are generally understood to be the result of an electric current produced when an electrically conductive fluid moves within the magnetic field – we can assume that there existed initially a weak "seed" as the magnetic field – with this current creating a permanent magnetic field (Faraday's Law). This theory is known as the hydromagnetic theory or magnetohydrodynamics. The earth has a core whose main component is thought to be an Fe–Ni alloy. We can conclude that the earth's magnetic field originated through the movement of the fluid in the core. Even today opinion remains divided over the cause of the fluid movement within the core. In the past it was thought that it was caused by thermal convection resulting from the energy released through the decay of the small amount of radioactive elements assumed to be present in the core. However, it is by no means clear whether or not the core, which consists mainly of an Fe–Ni alloy, actually contains significant amounts of such radioactive elements as U and K. Some researchers recently are of the opinion that the fluid core was formed in the initial stages of the earth's evolution, and that the thermal convection has been maintained as the core has gradually cooled off. Of late a theory that attributes the convection to the chemical differentiation of the fluid core has been gaining ground. Observations of seismic waves have given rise to the current view that a solid phase known as the inner core, with a radius of about 1000 km, exists in the middle of the core. As the core cooled, a solid phase would crystallize from the liquid phase. Since the crystallized solid phase would be heavier than the liquid phase, it would sink to the center of the core and form an inner core. According to this theory, the descent of the heavy phase would cause convection of the core fluid. This idea was first advocated by S.I. Braginsky in 1964. Thermodynamic considerations mean that this mechanical convection would be more efficient than thermal convection – with thermal convection the efficiency as a Carnot engine is limited, as is demonstrated by thermodynamics.

No matter what may be the cause behind the convection in the fluid core, the following two conditions are necessary in order for planets to have a permanent magnetic field: (i) a fluid core with good electrical conductivity; (ii) convection within the fluid core. If chemical separation is the sole mechanism behind the convection, the existence of a solid inner core is also a prerequisite.

The magnetic field properties of terrestrial planets are collated in Table 2.3. The table also shows the core characteristics that are of essential importance to the occurrence of a magnetic field. Based on this general discussion of the magnetic fields of planets, let us venture some conjectures as to the internal structure of each planet from a comparison of their magnetic fields (observed facts).

Table 2.3. Magnetic fields in terrestrial planets

	Radius (km)	Rotation period	Metallic core	Magnetic field
Mercury	2439	59 day	?	380 nT
Venus	6052	243 day	?	~ 30
Earth	6378	23.56 hour	Outer core: liquid Inner core: solid	~ 30 000
Mars	3397	24.37 hour	Solid core ?	~ 100

From the value of its moment of inertia, it has been concluded that Mars has a metallic core. On the other hand, no significant magnetic field has been confirmed. This suggests that the metallic core of Mars is solid, not fluid, or perhaps that even if it is fluid it is in a state in which it is difficult for convection to occur, for example, an inner core has not yet been formed. Perhaps Mars' slightly smaller size compared to the Earth has resulted in decisive differences in the thermal histories of the two planets, and in turn in the configuration of their cores.

Nor does Venus, which is roughly the same size as the Earth, have a significant magnetic field. The normal assumption would be that since it is about the same size as the Earth, it would have a similar thermal history and hence would have a metallic core just like the Earth. Since the rotating velocity of Venus is extremely slow, however, it is difficult to estimate its moment of inertia, and no confirmation of the existence of the metallic core has been made from dynamical astronomy. The lack of a general magnetic field on Venus may simply be due to the fact that it does not have a metallic core. Another conceivable cause, of course, is that Venus has a core, but no convection occurs.

A significant general magnetic field does exist on Mercury. The high density of Mercury suggests that this planet is rich in Fe–Ni, and so it seems appropriate for a metallic core to exist. The existence of the magnetic field backs up this likelihood. However, no moment of inertia data that actually prove the existence of a metallic core have been obtained.

As is obvious from the example above, comparative studies of the planets not only deepen our understanding of each planet, but also play a major role in contributing to our understanding of the fundamental physical and chemical processes – in the above example, the electromagnetic fluid theory related to the origin of magnetic fields – that regulate the origin and evolutionary processes of the planets.

e) How Old Is the Earth?

When was the earth formed? It was not until the 1960's that earth scientists became able to answer this basic question in earth science with any confidence. Up until then it had been almost impossible to form a reliable estimate even for the age of rocks, much less the age of the earth. The 1950's witnessed the first practical use of the radiometric dating method, and geochronologists vied to date "old" rocks around the world in an attempt to draw ever closer to the "age of the earth". In the 1950's and 1960's there was a rash of reports from sites in Africa, Australia and the North American continent of rocks aged 1000 or 2000 million years old, and it was an exhilarating time in which the record for the oldest rock value was revised each year. This excitement came to an end in the late 1960's, however, when S. Moorbath and his colleagues determined the age (approximately 3800 million years) of Isna metamorphic rock from eastern Greenland, and no older rocks have been reported since then. (Very recently W. Compston and his colleagues from Australia have reported that a few zircon grains contained in younger rocks in eastern Australia is approximately 4200 million years old). The source rock of Isna metamorphic rock is sedimentary rock, so this sedimentary rock must be more than 3800 million years old, and its source rock must be even older. It should be self-evident that the earth must be more than 3800 million years old. If we could obtain a sample that was representative of the whole earth, it would be possible in principle to date this sample directly and hence find the age of the earth.

However, the earth is a "living planet" which since its birth has undergone a continual process of material differentiation through igneous and volcanic activity, as well as mantle convection, and so on. Hence the material within our reach here on earth has all differentiated in greater or lesser degree from the original material constituting the earth, and bears little resemblance to the average composition of the earth. It is therefore impossible in principle to estimate the age of the earth directly by using samples of earth material. As stated in Section 2.3 a, what is regarded as the age of the earth nowadays is the age that has been concluded from the meteorite analogy, i.e., the age of meteorites = the age of the earth. Let us explain here in some detail why the meteorite analogy is justified as far as estimating the age of the earth is concerned.

With a few exceptions, all the solar material available to us for observation, including meteorites, earth material, and moon rocks, has an approximately uniform isotopic composition, and so it has been concluded that material within the solar system all broke away and separated from one and the same solar nebula. Several theoretical scenarios have been presented from an astrophysical viewpoint as to how the planets and parent bodies of meteorites formed from the solar nebula, and most of these scenarios hypothesize that planet formation was completed over a period of $10^7 - 10^8$ years. The dynamical astronomy scenario does not contradict the proposition that the earth and meteorites were formed more or less simultaneously. However, a scenario is merely a scenario, and whether or not matters actually occurred in line with the scenario is a different question. The simultaneous formation of the earth and meteorites must be verified from observed facts.

Direct proof of the simultaneous formation of the earth and meteorites can be found in the radiogenic ^{129}Xe (here referred to as $^{129}Xe*$) first detected in the Richardton chondrite. This discovery was made by J.H. Reynolds in 1960. ^{129}I has a half-life of 17 million years, and decays into ^{129}Xe*. The formation of ^{129}I within the solar system after the creation of the solar system can be ignored. We can consider all ^{129}I as having been formed in the nucleosynthesis that created the elements of the solar nebula. Owing to its short half-life, ^{129}I can be regarded as having decayed completely at a very early period in the history of the solar system – specifically, within a period equivalent to several half-lives. Reynolds succeeded in proving that a certain amount of the ^{129}Xe in the Richardton meteorite is the product of the radioactive decay of the ^{129}I that was contained in the meteorite. This demonstrates conclusively that the formation of meteorites occurred within a period equivalent to the half-life, i.e., a short period of several tens of millions of years, of ^{129}I after the completion of the nucleosynthesis.

It is inferred that some ^{129}I was also incorporated within the earth, and that the ^{129}Xe* formed through its disintegration has been degassed into the earth's atmosphere.

Collating these observed facts, we can conclude that both the meteorites and the earth were "formed" within just a few tens of millions of years of the end of the nucleosynthesis. It is difficult, however, to reach a concrete conclusion as to which process of the earth's evolution this "formation" corresponds. If we interpret it strictly from the experimental procedure (to be described below) used in indentifying ^{129}Xe*, it is the time when ^{129}Xe* became unable to escape from solid material, such as the earth and meteorites or the planetesimals that were their constituent material – in physical terms, the time at which the planetary material became a closed system as far as ^{129}Xe* is concerned. It is by no means clear at what stage in the growth of the earth ^{129}Xe* became unable to escape, or at what size we should regard the earth as having been born, i.e., the specific meaning of the age of the earth. Any further pursuit of this issue would degenerate into a semantic argument, so we will leave it here. When we combine the logical conclusion of the astrophysical scenario with the empirical fact of the existence of ^{129}Xe*, the virtually simultaneous birth of the earth and meteorites is an amply convincing conclusion. This means that we can regard the established age of meteorites of 4550 million years as the age at which the earth was formed.

Here we have examined the "simultaneous formation" that is one of the mainstays of the meteorite-earth analogy. The experiment proving the existence of ^{129}Xe* in meteorites, which provided the grounds for this conclusion, is based on an extremely elegant idea. Tracing the method by which this experiment was performed reveals some of the charm of scientific research – the beauty of the development and elegance of idea. This experimental method also contains the possibility that it can be applied in other fields also. Here we will look back in some detail over the experiment devised by P.M. Jeffrey and J.H. Reynolds.

Let us set up the problem again. Our aim is to demonstrate experimentally that some of the ^{129}Xe (the Xe existing in nature as a stable isotope consists of the nine isotopes ^{124}Xe, ^{126}Xe, ^{128}Xe, ^{129}Xe, ^{130}Xe, ^{131}Xe, ^{132}Xe, ^{134}Xe and ^{136}Xe) contained in meteorites and the earth is formed through the decay of the ^{129}I that formerly existed in meteorites and the earth. Jeffrey and Reynolds irradiated some meteorites with thermal neutrons. Neutron irradiation of the I (I consists only of the ^{127}I isotope) contained in tiny amounts within meteorites causes a ^{127}I (n,p) ^{128}I reaction so that it changes into ^{128}I. ^{128}I is a radioactive nucleus, and changes into the stable nucleus ^{128}Xe in a half-life of about 25 minutes. This means that meteorites that have been subjected to neutron irradiation contain the original ^{128}Xe and ^{129}Xe (including some ^{129}Xe*) and the ^{128}Xe* that was produced experimentally. Next let us conduct an experiment in which Xe is extracted from an irradiated sample. This can be done by heating the sample in a vacuum and expelling the gases. If the sample is completely melted all of the gases contained in it will be extracted. However, here we will follow Jeffrey and Reynolds by supposing that the temperature is gradually increased to 600 °C, 700 °C, 800 °C and so on, and held for a certain period (e.g., 30 min) at each temperature step, and that a gas sample is extracted at each temperature. This experimental method is known as the step-heating method. Next the samples obtained at each temperature are examined using a mass spectrometer, and the Xe isotopic ratios are measured. As noted in Section 2.2, it is convenient to express isotopes in the form of a ratio. For the sake of convenience, let us use ^{130}Xe as the common denominator. This enables us to obtain isotopic data for ^{129}Xe/^{130}Xe, ^{128}Xe/^{130}Xe ... and so on for each temperature step.

Amongst the Xe isotopic ratios obtained in the above experiment, let us focus particularly on the ^{129}Xe/^{130}Xe and ^{128}Xe/^{130}Xe ratios. We will plot the isotopic data obtained at each temperature, with ^{128}Xe/^{130}Xe on the X axis and ^{129}Xe/^{130}Xe on the Y axis. Figure 2.10 shows an example obtained for the Mundrabilla iron meteorite. There is an excellent correlation between the two isotopic ratios. Consider the meaning of this correlation. The isotopes ^{128}Xe and ^{129}Xe contained in meteorites consist of components formed from I (shown with an asterisk) through radioactive decay and components captured when the meteorite formed (shown with a subscript 0). Hence

$$\frac{^{129}\text{Xe}}{^{130}\text{Xe}} = \frac{1}{^{130}\text{Xe}}[(^{129}\text{Xe})_0 + (^{129}\text{Xe})^*]. \tag{2.20}$$

On the other hand, ^{130}Xe consists solely of a component that was captured when the meteorite formed, and so ^{130}Xe = $(^{130}\text{Xe})_0$. Here the amounts without subscripts can be observed empirically. Formally re-

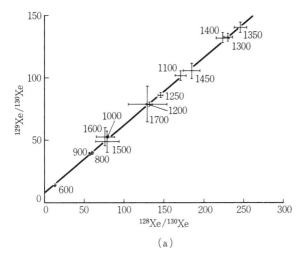

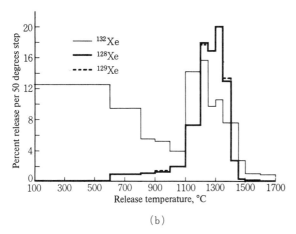

Fig. 2.10. Stepwise degassing of Xe from silicate inclusions separated from Mundrabilla iron meteorite which was irradiated by fast neutrons in a reactor. **a** Isotopic data obtained for gases extracted from the sample at various temperatures (shown by numbers) are plotted in a three isotope plot. The slope of the correlation line is proportional to $(^{129}Xe/^{128}Xe)^*$ or the I–Xe formation age. **b** Thermal release pattern of Xe. Note that iodine-induced ^{128}Xe ($^{128}I(n,r)^{128}Xe$) and radiogenic ^{129}Xe (*heavy black line*) decayed from ^{129}I show an almost identical release pattern which differs markedly from that for common ^{132}Xe (*thin line*). (After Niemeyer, 1976)

writing Eq. (2.20),

$$\frac{^{129}Xe}{^{130}Xe} = \frac{1}{^{130}Xe}\left[(^{129}Xe)_0 + \frac{(^{129}Xe)^*}{(^{128}Xe)^*} \times \{(^{128}Xe) - (^{128}Xe)_0\}\right]$$
$$= \left\{\left(\frac{^{129}Xe}{^{130}Xe}\right)_0 - \left(\frac{^{129}Xe}{^{128}Xe}\right)^* \left(\frac{^{128}Xe}{^{130}Xe}\right)_0\right\} + \left(\frac{^{129}Xe}{^{128}Xe}\right)^* \cdot \left(\frac{^{128}Xe}{^{130}Xe}\right).$$
(2.21)

The $^{128}Xe/^{130}Xe$ ratios obtained at each temperature fraction are plotted in Fig. 2.10a with $^{128}Xe/^{130}Xe$ on the X axis and $^{129}Xe/^{130}Xe$ on the Y axis. The neat linear correlation between x and y in Fig. 2.10a shows that the coefficient $(^{129}Xe/^{128}Xe)^*$ of the first and second terms in Eq. (2.21) does not depend on the extraction temperature, but is constant. This is perfectly understandable if we consider the process by which the step-wise degassing experiment is carried out. When step-heating is carried out, each component of the Xe isotope degasses in a different manner, mirroring the differences in the mechanism by which each component is captured in the meteorite crystal lattices and matrices. Meanwhile, similar isotopic components undergo similar degassing. This situation is shown in Fig. 2.10b. As a result, we can expect that isotopic ratios in which the numerator and denominator consist of the same component, such as $(^{129}Xe/^{130}Xe)_0$, $(^{128}Xe/^{130}Xe)_0$ and $(^{129}Xe/^{128}Xe)^*$, will have a constant value regardless of the temperature. The neat linear correlation in Fig. 2.10a bears out this expectation. Moreover, the value of $(^{129}Xe/^{128}Xe)^*$ can be found from the slope of the straight line in Fig. 2.10a. The fact that the slope is not 0 proves that $(^{129}Xe)^* \neq 0$, and hence that some $(^{129}Xe)^*$ formed through the decay of ^{129}I definitely did exist within this meteorite.

When a standard sample is also irradiated at the same time that the meteorite is subjected to neutron irradiation, the intensity of the neutron flux can be found, and from this it is possible to find the absolute value of $(^{129}Xe)^*$. Based on this value, we can estimate to a certain extent how long after the completion of the nucleosynthesis the formation of the parent body of the meteorite actually occurred, i.e., the formation interval of the meteorite.

As has been stated repeatedly, elemental differentiation means that there is virtually no possibility of earth material preserving intact the elemental composition at the time of the birth of the earth. Consequently, the step-heating method, which is extremely effective in elucidating the in situ decayed $^{129}Xe^*$ in meteorites, cannot be applied to earth material. Instead, a quite indirect method is used to verify the existence of $(^{129}Xe)^*$. The quantity of $(^{129}Xe)^*$ contained in different meteorites is extremely diverse. Certain kinds of meteorite (e.g., Nuvo

Urei) can be regarded as containing virtually no $(^{129}Xe)^*$. From the Xe isotopic composition of such meteorites it is possible to infer the Xe isotopic composition that does not contain $(^{129}Xe)^*$, i.e., the Xe isotopic composition as it was when formed through the nucleosynthesis, for now let us refer to this as primordial Xe. The specific procedure for finding the primordial Xe isotopic ratio is extremely complex and beyond the scope of this book. Here we will use this without going into an explanation of the actual procedure. The $^{129}Xe/^{130}Xe$ isotopic ratio of the Xe in the earth's atmosphere is as much as 7.6% higher than that of primordial Xe after correcting for a fractionation between air Xe and the primordial Xe. This difference can be regarded as the result of the $(^{129}Xe)^*$ formed through the decay of ^{129}I. Moreover, within the solar nebula $^{129}I/^{130}Xe \ll 1$, and $^{129}Xe^*$ can be virtually ignored. $^{129}Xe^*$ cannot be observed until the solid phase separates from the solar nebula and the solid element I is condensed in relatively greater quantities than rare gases, and $^{129}I/^{130}Xe \cong 1$. This guarantees that the $(^{129}Xe)^*$ observed on earth was derived through the decay of the ^{129}I that was contained in the earth material that separated as a solid phase.

From the above discussion we can conclude that meteorites and the earth had captured a certain amount of ^{129}I at the time of their formation, and this is further proof that the meteorites and earth were formed virtually simultaneously. We must be careful here, however, as to the specific meaning of "formation". Tracing back over the argument in this section, it will be self-evident that here "formation" means the time when the material forming the earth and the parent bodies of meteorites became a "closed system" as far as Xe and I are concerned. A discussion of Xe isotopic ratios alone is insufficient to reach any conclusion as to at what point in time in their growth, i.e., at what size, the earth and the parent bodies of meteorites became closed systems.

No definite estimate can be made of the age of the earth from earth material alone, but a number of constraints can be imposed. The Pb isotopic ratio is an example of this.

The first quantitative and meaningful discussions about the age of the earth were based on the isotopic ratio of Pb ores. This was the series of papers published from the late 1940's into the 1950's by F.G. Houtermans and A. Holmes, and later by C.C. Patterson et al. It is no exaggeration to describe the papers by Houtermans and Holmes as classics of isotope earth science. These papers were the first to demonstrate the importance and effectiveness of isotopic ratio data in throwing light on the origin and evolution of the earth.

All the earth material that is available to us consists of samples that were formed after having undergone differentiation after the earth was formed, and any records of the birth of the earth were wiped out in the process of this material differentiation. Thus these samples contain no information relevant to the birth of the earth, such as its age. By using earth samples, however, it is possible to trace the record of material differentiation. Houtermans and Holmes assumed that lead ores were formed by a process in which Pb was separated directly from the primeval earth (assuming a homogeneous earth which had undergone no material differentiation such as separation into mantle and core). It is obvious that the origin of Pb ores is not the result of such a simple mechanism, but let us accept this hypothesis for the time being. As will be explained in a later chapter, if it is possible to independently estimate the time at which the Pb ores were formed, we will be able to estimate the time at which the earth separated from the solar system, that is, the age of the earth. Using several Pb ores whose age had been estimated geologically, Houtermans and Holmes found the "age of the earth" as being 3000 to 5000 million years.

As is clear from current earth science, however, Pb ores separated not from a homogeneous earth that had remained unchanged since its birth, but from a differentiated earth after the separation of the core, mantle, and crust. The value found by Houtermans and Holmes is significant as the time of separation of the earth's crust and mantle, rather than as the age of the earth. Subsequently Patterson used the lead in deep sea sediment as the representative value of the crust (Patterson assumed that this deep ocean sediment is the result of crustal material being eroded and mixed and homogenized in the ocean on a global scale), and from a similar model to that of Houtermans and Holmes he found a value of 4500 million years as the age at which the earth's crust was formed.

The papers by Houtermans and Holmes and by Patterson are outstanding achievements worthy of the name classic, not only in the field of isotopic earth science, but also in earth science as a whole. Unfortunately, however, Houtermans, Holmes and also Patterson chose "Age of the Earth" as the title of their respective papers. The title of the papers and Patterson's apparent conformity of 4500 million years with the age of meteorites gave impetus to the general view that the age of the earth could be found from the Pb isotopic ratio within earth material. In actual fact, depending on the sample used, what can be estimated from the Pb isotopic ratio of earth material is the age at which the earth's crust and mantle separated, or the age at which the core and mantle separated. Since then many writers have attempted to estimate the age at which the mantle and core separated, using such samples as Pb ores, MORB–Pb (representative value of the differentiated mantle) and the Pb in deep ocean sediment (representative value of the earth's crust), and also using an elaborate Pb-isotope development model. These all have produced results ranging between 4200 and 4500 million years. The formation of the core is thought to be the earliest event on a global scale in the earth's history – many earth scientists believe that it occurred at the same time as the formation of the earth – and it is expected that it will show a value close to the age of the earth. As far as the age of the earth is concerned, this is the earliest event that can be differentiated from material available on earth.

3 Evolution of the Earth

3.1 The Driving Force Behind the Earth's Evolution

Mention was made in the previous chapter of the importance as a driving force behind the evolution of the "living earth" of the energy released through the nuclear disintegration of elements such as U, Th, and ^{40}K. In addition to the energy released through radioactive decay, however, the gravitational energy released in the process of the formation and evolution of the earth also plays an extremely important role as a driving force in the evolution of the earth. In the scenario of the earth's formation, for instance, we assumed that the countless number of planetesimals formed by breaking away from the solar nebula repeatedly collided with each other and combined to grow into a large planet (earth). As the earth gradually grew in size and its mass increased, its gravity also increased. Thus planetesimals that collided and combined after the earth had become quite large would have been drawn to the earth by a larger gravitational force, and would have released a greater amount of gravitational energy when they collided. Some of this energy would have laid waste to the surface of the earth, and some would have been stored in the earth as thermal energy or released into space. If the planetesimals collided repeatedly in a short period of time, there would have been little time for the resulting energy to escape from the surface of the earth, and further energy would have been added through the subsequent impact of other planetesimals, and so the energy would be stored in the earth's interior more efficiently. The amount of energy stored in the earth's interior when the earth was formed is closely related to the duration of the formation of the earth. The upper limit of the energy provided to the earth at the time of its formation – let us refer to this as the "earth accretion energy" – corresponds to the potential energy if the material comprising the earth were dispersed to infinity. Simple calculations show that this value is approximately 2.5×10^{32} J, more than sufficient to melt the earth completely. All of this energy would not be stored in the earth, since a considerable amount would escape from the earth's surface into space as infrared radiation. The amount of energy stored at the time of the earth's formation would have been determined by the balance between the energy released from the earth and the rate of accretion of the earth. Recent estimates put the amount of energy contained in the earth's interior at about one-thirtieth of the total accretion energy, assuming that it took the earth 10^7 years to accrete.

The gravitational energy released after the formation of the earth as the material within the earth is redistributed – this is the actual process of evolution – must also be considered as a major energy source promoting the evolution of the earth. Let us suppose for now that the earth was born as a planet that was homogeneous from its center out to its surface, and that the separation of the mantle and core commenced from this homogeneous earth. Specifically, let us suppose that iron separated from the silicate materials in the earth's interior, and that the weight of the iron caused it to sink within the silicates until finally a core was formed. Here we will not go into the question of how the iron separated and managed to collect in the center to form a core. If the heavy iron components sink within the earth's gravity field, potential energy would be released. This energy has been estimated at approximately 1×10^{31} J.

These two kinds of energy are the result of the release of gravitational energy, but the energy released by radioactive elements is also important. This is the heat produced by the kinetic energy of particles emitted at the time of nuclear disintegration. When radioactive elements undergo nuclear disintegration they emit particles such as electrons and particles. In due course these particles collide with the other atoms of which the material is composed, and eventually come to a stop within the material. Their kinetic energy then turns into heat and is stored within the material. Judging from their abundance in the earth's interior, the four isotopes (Rb is shown just for a comparison) shown in Table 3.1 are the only radioactive elements that are important heat sources for the earth. All of these isotopes have a long half-life of hundreds of millions of years. Table 3.1 also shows the energy released when these elements decay. The amount of radioactive elements within the earth decreases gradually over time owing to radioactive decay. This means that the

Table 3.1. Heat generation of radioactive elements

	(10^{-13} J/atom)	(W/g)
^{238}U	75.9	0.94
^{235}U	72.4	5.7
U		0.97
^{232}Th	63.7	0.26
^{40}K	1.14	0.288
K		3.43×10^{-5}
^{87}Rb	0.07	2.1×10^{-3}
Rb		5.8×10^{-5}

thermal energy produced through radioactive decay in the earth's interior has also decreased exponentially since the birth of the earth. The thermal energy of radioactive decay origin that has been released throughout the whole 4500 million years of the history of the earth can be calculated as appproximately 1×10^{31} J.

Another source of heat for the earth is radiative energy from the sun. The sun is continually releasing into space the massive energy produced through nuclear reactions inside the sun. This energy is currently estimated at about 3.83×10^{26} J per second. In the vicinity of the earth's orbit, therefore, the solar energy received per unit area (1 cm^2) of the earth which is perpendicular to the sun's rays is 0.14 J per second, or 8.4 J per minute. This is normally known as the solar constant. Hence the whole earth receives radiative heat of 0.14 J × (cross-section of the earth) = 1.8×10^{17} J per second. Solar energy is the energy source that supports biological activity, and it is the decisive factor governing phenomena on the surface of the earth, such as weather conditions and climate. However, it can be virtually ignored as an energy behind the evolution of the earth. Practically all of the radiative energy that reaches the surface of the earth from the sun is eventually reflected back into space as infrared radiation, and is not stored within the earth. This is illustrated in Fig. 3.1. Thirty percent of the radiation from the sun is reflected by the surface of the earth and by clouds and the atmosphere. The energy that reaches the surface of the earth and is provided to the earth's interior as heat accounts for no more than about 50% of the total

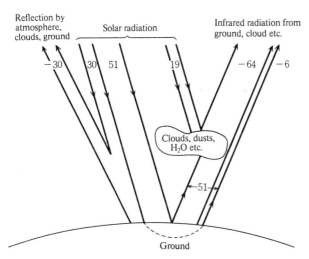

Fig. 3.1. Incident solar radiation is either reflected (about 30%) or re-radiated as infra-red light (about 70%) from the earth and there is no storage of the solar energy inside the earth

Table 3.2. Energy sources for earth evolution

(i)	Accretion energy	$E \cong 2.5 \times 10^{32}$ J
(ii)	Core differentiation	$E \cong 10^{31}$ J
(iii)	Radioactive decay (for 4.5 billion years)	$E \cong 10^{31}$ J

energy radiated by the sun. The solar energy that reaches the earth is used up as the driving force for the atmosphere and oceans or to support biological activity, but eventually it is again released into space as infrared radiation. Although the energy radiated by the sun can be virtually ignored as an internal source of energy behind the evolution of the earth, it has an important influence on climatic changes. Some theories attribute the ice age to changes in the energy emitted by the sun.

Table 3.2 shows a comparison of the three basic sources of energy behind the evolution of the earth. The amount of earth accretion energy varies greatly depending on the assumed duration of the formation of the earth, and this is the most indefinite figure. Even taking this indefiniteness into account, however, it is worthy of note that the three kinds of energy are all roughly equivalent in magnitude. On the other hand, there is a great difference in how they affect the evolution of the earth. The accretion energy that was only available during the formation of the earth was the most important influence in the initial stages of the earth's evolution. By contrast, radioactive isotopes with a long half-life have had a steady influence throughout the whole history of the earth. Volcanic activity and continent-building activity throughout geological times are caused mainly by the radioactive decay energy accumulated in the earth's interior. As for the energy produced by the redistribution of material within the earth – in particular, the energy produced by the separation of the mantle and core – the core separated in the very early stages of the earth's history, and so this energy would have had the most effect in the initial stages of the earth's evolution. The energy produced by material redistribution consists not only of that produced by the core-mantle separation, but also the energy released by the separation of the inner core. This energy is now regarded as an important source of energy in maintaining the earth's magnetic field. This will be discussed in further detail in the next chapter in the section on the geomagnetic field.

3.2 Composition of the Earth – The Meteorite Analogy

How much of the U, Th, and K that have acted as a driving force throughout the evolution of the earth has the earth contained? What is the composition of the earth as far as other elements are concerned? Numerous researchers have estimated the elemental composition of the earth, as it is one of the most fundamental issues in earth science. However, our present knowledge of the earth's chemical composition is far from complete. Logically speaking, we should be able to estimate the chemical composition of the earth by selecting and analyzing a sample representing the whole earth, but the problem lies in obtaining such a representative sample. No matter over how wide a range rock samples may be collected, the samples that we can actually lay our hands on are insignificant fragments from the viewpoint of the earth as a whole. The deepest bore hole that has been drilled so far is one on the Kola Peninsula in the USSR, reaching 13 km down below the surface. If the earth is likened to an apple, however, this does not even scratch the surface of the skin. Mantle fragments (referred to as mantle xenoliths) that are occasionally brought to the surface of the earth together with volcanic rocks are the sole source of information on the composition of the earth at depths beyond this level. Even these mantle xenoliths and volcanic rocks were derived from depths of 100–200 km at the most, and so they are far from providing a lead to the average composition of the mantle. In order to determine the average chemical composition of the whole earth we cannot rely solely on the samples available to us directly, and are forced to depend mainly on indirect methods. As mentioned in Chapter 2, the most effective of these methods is the meteorite analogy.

In order to understand the meteorite analogy, let us look back again over the solar system-planet formation scenario. The uniformity (with a few exceptions) of the isotopic compositions of material found in the solar system indicates that the solar nebula from which the planets formed was considerably homogenized. Astrophysical theory, which deals with the movement of the solar nebular gas, also predicts that the solar nebula was well-mixed and homogenized. We can conclude, therefore, that the planets and meteorite parent bodies formed from the solar nebula also had similar chemical compositions. This is the major premise of the meteorite analogy. It can easily be imagined, however, that the high volatility of volatile elements means that in high-temperature regions closer to the sun they would not be able to condense, and considerable fractionation would occur, thus destroying this uniformity. It has been concluded that on Mercury, the planet closest to the sun, volatile elements cannot condense because of the high temperature, and so metal elements are the main constituent elements of this planet, thus resulting in the high density that has been observed. On the other hand, nonvolatile elements, which can condense at high temperatures, condense on all planets, and no differences caused by different orbits would appear. Consequently, we can expect all solar material, such as planets and the parent bodies of meteorites, to have a highly similar composition as far as nonvolatile elements are concerned. In the case of meteorites such as carbonaceous chondrites, which became "dead" objects as soon as they were born, and have undergone almost no thermal or chemical differentiation up to the present, the composition of their parent body can be estimated from an analysis of a very tiny part of the parent body, i.e., meteorites that have fallen to earth. This also provides a clue as to the chemical composition of the solar nebula from which the parent body was created. This is the theory on which the meteorite analogy is based.

The meteorite analogy, which is the guiding principle in estimating the chemical composition of the earth, has been advocated mainly on the premise of the uniformity (at least for nonvolatile elements) of the solar nebula. How does the meteorite analogy rate on the basis of observed results from the earth itself? From the 1960's onwards, systematic heat flow measurements have been carried out on the ocean floor. The results have revealed that the heat flow from the surface of the earth is approximately 75 mW m^{-2} (1.8 × 10^{-6} cal cm^{-2} s^{-1}). It has been pointed out that the internal heat source necessary to create this heat flow is roughly equal to the amount produced by the radioactive elements contained in chondrites, and this was viewed at one time as powerful proof of the meteorite analogy, although it is now regarded as a fortuitous coincidence. If the earth's core consists mainly of Fe, Fe would account for about 30 percent of the earth's mass, and this is more or less equivalent to the abundance of Fe in chondrites. The material remaining if Fe were removed from chondrites closely resembles $MgSiO_3$ and is akin to pyroxene, which is one of the main minerals constituting the present mantle. This gives convincing evidence of the similarity of the chemical compositions of the earth and meteorites.

Hence it seems reasonable on the whole to regard the meteorite analogy as being valid in the case of nonvolatile elements. For volatile elements, however, the meteorite analogy needs considerable revision. One example is the case of Rb. The paper published by P. Gast in 1960 was one of the first papers to criticize the meteorite analogy, and is viewed as a classic in earth science. Comparing the $^{87}Sr/^{86}Sr$ ratios of meteorites and earth material, Gast pointed out that the Rb/Sr ratio of the earth's mantle is an order of magnitude smaller than that of meteorites. The general view is that this is because Rb, which is more volatile than Sr, was lost from the earth (some scientists are of the opinion that it was incorporated in the earth's core). If this is the case we can expect K, which like Rb is also highly volatile, to be deficient in the earth compared to its abundance in meteorites.

Based on the meteorite analogy, Table 3.3 shows the earth's elemental composition, taking into account such geophysical data as the heat flow and density distribution of the earth. This work was carried out by R. Ganapathy and E. Anders. In Table 3.3 we also show the

Composition of the Earth – The Meteorite Analogy   61

Table 3.3. Elemental abundances in the Earth and moon. (Ganapathy and Anders 1974)

Element	Moon	Earth	Element	Moon	Earth
H	2.2	78	Ru	4.9	1.48
He^4	2100	74 000	Rh	1.05	0.32
Li	8.7	2.7	Pd	0.25	1.00
Be ppb	186	56	Ag ppb	9.6	80
B ppb	13	470	Cd ppb	0.58	21
C	9.9	350	In ppb	0.075	2.7
N	0.26	9.1	Sn	0.085	0.71
O%	41.42	28.50	Sb ppb	7.6	64
F	30	53	Te	0.20	0.94
Ne^{20}	7	250	I ppb	0.48	17
Na	900	1580	Xe^{132}	0.13	4.8
Mg%	17.37	13.21	Cs ppb	33	59
Al%	5.83	1.77	Ba	16.8	5.1
Si%	18.62	14.34	La	1.57	0.48
P	538	2150	Ce	4.2	1.28
S%	0.39	1.84	Pr	0.53	0.162
Cl	0.70	25	Nd	2.9	0.87
Ar^{36}	37	1330	Sm	0.86	0.26
K	96	170	Eu	0.33	0.100
Ca%	6.37	1.93	Gd	118	0.37
Sc	40	12.1	Tb	0.22	0.067
Ti	3380	1030	Dy	1.49	0.45
V	340	103	Ho	0.33	0.101
Cr	1200	4780	Er	0.96	0.29
Mn	330	590	Tm	0.145	0.044
Fe%	9.00	35.87	Yb	0.95	0.29
Co	240	940	Lu	0.160	0.049
Ni%	0.51	2.04	Hf	0.95	0.29
Cu	6.9	57	Ta ppb	96	29
Zn	19.9	93	W	0.75	0.250
Ga	0.66	5.5	Re ppb	250	76
Ge	1.66	13.8	Os	3.6	1.10
As	0.90	3.6	Ir	3.5	1.06
Se	1.30	6.1	Pt	6.9	2.1
Br ppb	3.8	134	Au	0.072	0.29
Kr^{84}	0.18	6.6	Hg ppb	0.28	9.9
Rb	0.33	0.58	Tl ppb	0.136	4.9
Sr	60	18.2	Pb^{204} ppb	0.055	1.97
Y	10.9	3.29	Bi ppb	0.104	3.7
Zr	65	19.7	Th ppb	210	65
Nb	3.3	1.00	U ppb	59	18
Mo	9.8	2.96			

elemental composition of the moon estimated on the basis of the meteorite analogy.

3.3 The Layered Structure of the Earth

Methods based on analyses of seismic waves remain the most effective means of investigating the internal structure of the earth, especially such physical conditions as its elasticity constant or density distribution. Earthquake-caused disturbances inside the earth are transmitted throughout the earth's interior as elastic waves. The manner in which these elastic waves are transmitted, i.e., their velocity, is determined by the density of the material and its elastic constant. Thus an analysis of the records of several observation stations around the world will reveal the velocity of the seismic waves, making it possible to estimate the elastic constants and density of the part of the earth's interior through which the seismic waves have passed. Taking into consideration the results of high pressure physics, which investigates the physical state of material under high temperatures and high pressure, it is also possible to estimate the material composition of the earth's interior. From the results of such studies it has been concluded that the earth consists of three basic structures – a core (distance from the center ranging from 0–3450 km), the surrounding mantle (3450–6350 km) and the uppermost layer of the earth's crust, which corresponds to a surface skin about 30 km thick. The core has a density of more than 10 g cm^{-3}. Fe is the only element that would have such a high density under the pressure in the center of the earth and that, judging from its solar abundance, is sufficient in quantity to form a core with a radius of 3500 km. From this argument we can safely conclude that the core of the earth has an elemental composition consisting mainly of Fe. The manner in which seismic waves are transmitted reveals that the core is divided into a solid inner core with a radius of approximately 1300 km and a liquid outer core surrounding that. The earth's magnetic field is thought to be the result of the fluid movement of the liquid outer core. The geomagnetic field will be discussed in detail in a later chapter.

It is virtually certain that the earth's core consists of a metal whose main component is Fe, but further examination reveals that the density of the core (under the high temperature and pressure conditions in the earth's interior) is approximately $1-2 \text{ g cm}^{-3}$ less than would be expected in the case of pure Fe. This suggests strongly that the core also includes elements that are 10–20% lighter than Fe. Various theories have been put forward suggesting that these light elements may be O, S, or Si, but no final conclusion has been reached.

The mantle accounts for about 83% of the total volume of the earth. The density of the mantle is $4-6 \text{ g cm}^{-3}$, or half that of the core, which consists of metal elements. Analyses of mantle xenoliths, which are the only direct samples available to us, analyses of seismic waves, and solar abundances have led to the conclusion that the material composition of the mantle is similar to that of olivine [$(Mg,Fe)_2SiO_4$] and pyroxene [$(Mg,Fe)SiO_3$]. The velocity at which seismic waves are transmitted reveals that the mantle is divided into three layers – the upper mantle (from the surface of the earth to a depth of about 400 km), the lower mantle (from a depth of 700 km to the core) and a transitional layer between these two layers. It is not yet clear whether or not the layered structure within the mantle is the result of differences in the component material of each layer, but the results of high-pressure physics studies in recent years strongly suggest that it is primarily the result not of such differences, but of the phase transition of crystals under high pressure.

Despite the fact that at less than 0.1 percent the earth's crust accounts for a mere fraction of the earth's mass, it is the layer which is most closely connected with the human race. It is possible to estimate the composition of the earth's crust on the basis of the analyzed values of samples that have been collected directly. The rocks used as samples must be representative of the earth's crust. The analytical data compiled by F.W. Clark and H.S. Washington on an enormous number of crustal rocks and the analysis of glacial sedimentary rocks by V.M. Goldschmidt are examples of such attempts to estimate the composition of the earth's crust. When glaciers move they scrape off and mix in surface rocks over a wide range, so Goldschmidt expected that these rocks would have a value close to the average composition of the earth's crust. Despite the fact that the number of samples and the sites at which the samples were collected differ greatly in the above two cases, the results are highly consistent. Both results also agree in general with the average crustal composition subsequently estimated by A. Poldervarrt (1955) and S.R. Taylor (1977) on the basis of more precise and abundant analytical values. Table 3.4 shows the results found by Taylor in one of the latest such attempts. It is estimated that the composition of the earth's crust resembles a mixture of basalt and granite in a ratio of approximately 3:1.

Table 3.4. Chemical composition of the crust. (After Taylor 1977)

	wt%
SiO_2	58.0
TiO_2	0.8
Al_2O_3	18.0
Fe_2O_3, FeO	7.5
MgO	3.5
CaO	7.5
Na_2O	3.5
K_2O	1.5

3.4 Formation of the Layered Structure

When and how was this layered structure of the earth formed? The simplest explanation would be to assume that when the earth was formed through the collision and accumulation of planetesimals, at first metal planetesimals consisting mainly of Fe accreted, followed by planetesimals consisting mainly of silicates and having a similar composition to that of the present mantle, followed finally by the accretion of crustal material. According to this hypothesis, known as the heterogeneous accretion model, the earth was formed by different materials accreting at different stages in the earth's development. This hypothesis avoids the problem of how the core and layered structure were formed after the formation of the earth, and is an extremely direct hypothesis. Dynamically, however, there is no rational explanation for how Fe and other metals could accrete first, followed by silicate materials. Since the planetesimals formed virtually simultaneously (probably within a period of 10^5 years) owing to the gravitational instability of the primitive solar nebula (see Chap. 2), and since the primitive solar nebula was quite well homogenized, we can expect rather that the material composition of planetesimals was homogeneous. Hence it seems more reasonable to consider that from start to finish the accretion of the earth occurred through the accretion of planetesimals with a more or less homogeneous composition. This is the viewpoint of the homogeneous accretion hypothesis. Consequently, the material composition of the newly accreted earth would have consisted mainly of silicates, similar to meteorites.

In line with the homogeneous accretion hypothesis, let us suppose that the new-born earth had a virtually homogeneous composition from its center out to its surface layer. We must explain at what point in the earth's evolution and by what process the current layered structure of core, mantle, and crust formed. An Fe core could not separate from a more or less homogeneous earth unless the earth was at a considerably high temperature and the silicates were in a virtually molten state. Accordingly, the question of the time at which the core separated boils down to a discussion of the thermal evolution of the earth, and specifically to the question of when did a major part of the earth reach a temperature close to melting point.

The most important factor governing the thermal history of the earth is the earth's heat sources. Already in Section 3.1 we have listed the most important heat sources as (i) the gravitational energy released when the earth accreted, (ii) the energy released when the material within the earth redistributed and (iii) the energy released through the nuclear disintegration of radioactive elements. It was also stated that these sources were roughly the same in magnitude, and that in the very early stages of the earth's evolution (i) played the major role, whereas throughout the rest of the earth's history the energy released through radioactive decay (iii) has had a steady effect. It is fully possible that if the earth accreted over a short period of time the accretion energy in (i) would have had a larger value than the other two sources of heat. In Chapter 2 we pointed out that once the core began to accrete, its formation, i.e., the release of gravitational energy through the subsidence of heavy metal elements, may have had a synergistic effect whereby the temperature rose, thus accelerating the formation of the core. Hence we can assume that the core formed in the very early stages of the earth's evolution, and that its formation was completed quite rapidly. As will be discussed in a later section, this conclusion has also been reached from Pb isotopic compositions. Before moving on to a discussion of the time of core formation based on Pb isotopic ratios, let us here add a few comments about the separation of the core.

Planetesimals, which are the basic constituting block in the formation of the earth, have a more or less homogeneous composition, at least as far as nonvolatile elements are concerned – their composition resembles that of primitive meteorites such as carbonaceous chondrites. The main constituent minerals of chondrites are an Fe-Ni alloy, troilite (FeS) and olivine [$(Mg,Fe)_2SiO_4$], as well as such silicates as orthopyroxene [$(Mg,Fe)SiO_3$] and plagioclase ($NaAlSi_3O_8 - CaAl_2Si_2O_8$). Hence we can assume that the newly accreted earth had a composition similar to this. The minerals constituting chondrites are all crystals of several millimeters or less, and after accretion the earth also consisted of a collection of these small crystals. Hence we return to the question of how tiny crystals (including an Fe–Ni alloy) smaller than a millimeter in size were able to form a large metal core with a radius as large as 3500 km. We are far from a final conclusion regarding this problem. Nevertheless, several fundamental processes involved in the formation of the core have been discussed in quantitative and persuasive terms. Here we will introduce the core formation scenario proposed by K. Sasaki and K. Nakazawa.

If the earth completed its accretion within the solar nebula, its outer layer would have reached quite a high temperature owing to the insulatory (or blanket) effect of the earth's atmosphere. According to calculations by Sasaki and Nakazawa, when the earth reached a stage where its mass was about one-sixth of the present mass, its surface temperature would have exceeded the melting point of silicates, and the upper layer of the earth would have begun to melt. As accretion proceeded, the temperature of the upper layer would continue to rise. Meanwhile, in the upper layer iron would separate from the melted silicates and begin to subside, but it would not sink right down to the center of the earth. This is because at the time of its formation the outer layers of the earth would have been hotter than the interior (owing to accretion energy), so the iron that had sunk relatively rapidly at the upper layers of the earth would have had difficulty in sinking in the lower layers because the viscosity of the surrounding silicates would increase as the temperature fell. Eventually the iron would have accreted around the relatively low-temperature region in the center of the earth – here the low temperature means that the silicates would have been highly viscous, and so the iron could sink no further. Since the heavy iron layer had settled on top of the light silicates, however, it would be gravitationally unstable, and the outer iron and inner silicates would undergo a catastrophic overturn (known as Rayleigh-Taylor gravitational instability) and the silicates would be forced outside, thus forming a core of iron in the center. The main points of this theory proposed by Sasaki and Nakazawa had already been discussed qualitatively by W.M. Elsasser in the 1960's.

Here we have reviewed a scenario that commenced with a homogeneous earth in which the core later separated. Despite the fact that the homogeneous accretion of the earth, which was the basic premise of this scenario, is dynamically valid, some researchers have cast doubt on the hypothesis that the formation of the Fe core proceeded in the form of separation of the Fe phase from the silicates. Their objections are based on the amount of siderophile elements currently contained in the mantle.

Siderophile elements are the elements belonging to the eighth group on the periodic table, mainly Ni, Co, Re, Os, and Pt, and they have a very strong affinity with Fe. For instance, it is known that when liquid Fe and silicates (containing some Ni) are made to coexist under an equilibrium condition in laboratory experiments, siderophile elements are over 1000 times more concentrated in the Fe phase than in the silicates. Hence if core formation proceeded in accordance with the scenario described above, it is highly likely that the siderophile elements would shift from the silicates to the subsided Fe, and in the long run most of the siderophile elements would be incorporated in the Fe core. If the silicate crystals composing the earth in its early stages were actually tiny crystal particles a millimeter or less in size, as is generally thought, then we can expect that under high temperatures the diffusion would be fast enough so that the siderophile elements would escape from the silicate crystals and be absorbed in the surrounding Fe phase. It is estimated that the present mantle contains about 2000 pm of Ni, one of the most typical siderophile elements. This means that approximately 8×10^{24} g are contained in the mantle as a whole. Meanwhile, if the relative abundances of such nonvolatile elements as Fe, Ni, and Si are the same as their solar abundances, the earth would contain approximately 2% of Ni, an amount equivalent to 1.2×10^{26} g for the whole earth. Therefore the ratio between the Ni content of the mantle and that of the core is approximately $8 \times 10^{24}/4.1 \times 10^{27} : (1.2 - 0.08) \times 10^{26}/1.9 \times 10^{27} = 0.033 : 1$ (denominators on the left-hand side denote the mass of the mantle and core respectively). The ratio in which Ni is distributed between silicates and Fe in laboratory experiments is approximately $10^{-3}:1$, and is known as the distribution coefficient. Judging from the experimental value for the distribution coefficient of Ni, therefore, if the distribution of Ni occurred under an equilibrium condition we can expect Ni to be approximately 1000 times more concentrated in the metal phase, which consists mainly of Fe, than in the silicate phase, i.e., the mantle. As shown above, however, in actual fact the mantle contains almost 2000 ppm of Ni, and the concentration of Ni in the core is at the most only about 30 times that in the mantle. Despite the uncertainty, e.g., the effect of pressure and temperature, of applying the experimental distribution coefficient value to the earth, the vast discrepancy between the experimental value and the actual value is definitely significant. Focusing on this discrepancy, some researchers have claimed that the formation of the earth's core cannot be explained as the separation of the silicate and Fe phases under an equilibrium condition, and that the homogeneous accretion theory, which is the basic premise on which the silicates-Fe separation theory is based, is incorrect.

It is true that it is hard to reconcile the existence of excess Ni in the mantle with the simple silicates-Fe separation theory. Rejecting the homogeneous accretion hypothesis and advocating a heterogeneous accretion hypothesis, however, brings one up against even greater difficulties. One of these is the fact that an appropriate dynamic model cannot be formulated to explain this heterogeneous accretion. In order to avoid these difficulties, it has been proposed that Ni was incompletely partitioned into the Fe phase because the separation into the Fe and silicate phases occurred in a disequilibrium condition. Alternatively, some scientists have proposed that some planetesimals that fell during the final stages of the earth's accretion came to rest on the surface layer of the earth, where owing to the relatively low temperature their Fe and Ni were preserved intact without undergoing Fe-silicate separation, and that they did not assimilate into the mantle until after the formation of the core. Taking into account the fact that the distribution of Ni between the Fe and silicate phases depends on the partial pressure of oxygen (P_{O_2}), there are also some researchers who claim that under the P_{O_2} conditions in the interior of the primeval earth a considerable proportion of Ni would have been partitioned in the silicate phase when the Fe and silicate phases were in equilibrium. At all events, the general outline of homogeneous accretion of the earth followed by separation of the Fe core seems reasonable. The Pb isotopic data to be discussed below do not contradict the interpretation that the Fe core separated from a homogeneously accreted earth. Pb isotopic data also suggest strongly that the separation of the Fe core occurred within a few hundred million years of the birth of the earth. Let us discuss these data below.

3.5 The Time of Core Formation, Based on Pb Isotopic Ratio Data

The Pb in nature consists of four isotopes, ^{204}Pb, ^{206}Pb, ^{207}Pb, and ^{208}Pb. With the exception of ^{204}Pb, these contain components formed through radioactive decay from U or Th. Thus Pb consists of components formed through nucleosynthesis and radiogenic components that were formed after the birth of the earth through the disintegration of the U and Th contained in the earth. Marking the components formed through nucleosynthesis with a subscript o and the radiogenic isotopes with a superscript *, the Pb isotopes observed today can be written respectively as

^{204}Pb = $(^{204}$Pb$)_0$,
^{206}Pb = $(^{206}$Pb$)_0$ + $(^{206}$Pb$)^*$,
^{207}Pb = $(^{207}$Pb$)_0$ + $(^{207}$Pb$)^*$
and ^{208}Pb = $(^{208}$Pb$)_0$ + $(^{208}$Pb$)^*$.

Let us next consider the evolution of Pb isotopic ratios within the earth's interior. As discussed in the previous section, suppose that the primeval earth (age of the earth = t_0) was homogeneous in composition. Suppose then that t_c years ago the Fe core separated from the homogeneous earth. The changes in the Pb isotopic ratios at that time would be as shown in the following equations. As explained in Chapter 2, here all isotopes are expressed in the form of isotopic ratios in order to simplify the comparison with experimental values. To simplify matters, here we will confine our discussion to the two Pb isotopic ratios ^{206}Pb/^{204}Pb and ^{207}Pb/^{204}Pb formed through the decay of U. We will use ^{204}Pb, which does not change over time, as the denominator. Taking into account the fact that $(^{238}$U/^{235}U) = 137.8 (current value), the Pb isotopic ratios at present (t = 0) are as follows:

$$^{206}\text{Pb}/^{204}\text{Pb} = (^{206}\text{Pb}/^{204}\text{Pb})_0 + \mu_E(e^{\lambda t_0} - e^{\lambda t_c}) + \mu_E(e^{\lambda t_c} - 1)$$

$$^{207}\text{Pb}/^{204}\text{Pb} = (^{207}\text{Pb}/^{204}\text{Pb})_0 + \frac{\mu_E}{137.8}(e^{\lambda' t_0} - e^{\lambda' t_c})$$

$$+ \frac{\mu_E}{137.8}(e^{\lambda' t_c} - 1), \tag{3.1}$$

where λ and λ' denote the decay constants for ^{238}U and ^{235}U, and μ_E, μ_M denote the values $(^{238}$U/^{204}Pb) for the total earth and the mantle respectively.

The right-hand sides of both of these equations consist of three terms, with the first terms showing the value at the completion of the nucleosynthesis. Since U has a long half-life, we can regard the end of the nucleosynthesis and the birth of the earth as having occurred at the same time, as far as the changes in the Pb isotopic ratios are concerned. The second terms correspond to the contribution made by the Pb isotopes that have been added through the disintegration of U during the period from the birth of the earth up until the separation of the core. The third terms correspond to the Pb added since the formation of the core. Specifically, the second terms correspond to the Pb that disintegrated during the time $t_0 - t_c$ on the homogeneous earth $[\mu_E \equiv (^{238}U/^{204}Pb)_E]$, and the third terms correspond to the Pb formed in the remaining mantle $[\mu_M \equiv (^{238}U/^{204}Pb)_M]$ from the time of core separation t_c up until the present. In this discussion all quantities without subscripts or superscripts express present values.

Changes in the Pb isotopic ratio are determined by the amount of U, or more precisely by $^{238}U/^{204}Pb$ ($\equiv \mu$). The equations in (3.1) were deduced by assuming that when the core separated from the homogeneous earth the value of U/Pb changed from μ_E to μ_M. What effect did core separation have on U/Pb? This can be discussed in terms of the distribution of U and Pb between the Fe phase and the silicate phase as explained in the previous section. So far there have been very few experiments on the distribution of U and Pb between the Fe and silicate phases. One of the few examples are the experiments carried out by V.M. Oversby and A.E. Ringwood, the results of which show that the Fe phase contains almost no U. On the other hand, Pb is distributed more selectively in the Fe phase than in the silicate phase. The distribution coefficient $[D = (Pb)_{Fe}/(Pb)_{silicate}]$ thus estimated is about 2.5. Hence if this experimental distribution coefficient can be applied as is to the distribution of Pb between the Fe phase and silicate phase in the earth's interior, it will mean that Pb is approximately 2.5 times more concentrated in the core than in the mantle.

Let us return to the equations in (3.1). The Pb isotopic ratios on the left-hand side in this equation are the present mantle values, and can be estimated from igneous rocks of mantle origin. The first terms on the right-hand side, $(^{206}Pb/^{204}Pb)_0$ and $(^{207}Pb/^{204}Pb)_0$, are primordial lead, and are found from the analyzed values of the troilite (FeS) phase in iron meteorites. Whereas troilite contains some Pb, it contains almost no U or Th. Hence its Pb isotopic ratios can be regarded as having remained virtually unchanged ever since the formation of iron meteorites. Thus the isotopic ratios of the lead in troilite can be regarded as the isotopic ratios of primordial lead. So t_c and μ_E and μ_M remain as the unknown quantities in the equations in (3.1). Let us now consider the Pb distribution at the time of core separation. The distribution coefficient D_{Pb} has been defined as the ratio between the Pb concentration in the Fe phase (core) and that in the silicate phase (mantle), i.e., $D_{Pb} \equiv (Pb)_C/(Pb)_M$. Suppose that U was distributed only in the mantle and did not enter the core. From a simple mass balance calculation we obtain

$$\mu_E \equiv \frac{(^{238}U)_E}{(^{204}Pb)_E} = \frac{(^{238}U)_M \cdot (1 - \beta)}{(^{204}Pb)_M(1 - \beta) + (^{204}Pb)_M \cdot \beta \cdot D_{Pb}}$$

$$= \mu_M \cdot \frac{(1 - \beta)}{(1 - \beta) + \beta \cdot D_{Pb}}. \quad (3.2)$$

β is the ratio between the volume of the core and that of the mantle. Eliminating μ_E and μ_M from Eqs. (3.1) and (3.2) produces the relation between t_c and D_{Pb}. Figure 3.2 plots this relationship with D_{Pb} on the vertical axis and t_c on the horizontal axis. If the experimental value (approximately 2.5) for D_{Pb} holds good for the earth's interior, then the time t_c of core separation corresponding to $D_{Pb} = 2.5$ can be calculated from Pb in this figure as being approximately 4200 million years.

Figure 3.2 was drawn up by R. Vollmer, but the concepts expressed in the equations in (3.1) were first presented by Oversby and Ringwood. The values chosen by Vollmer for the left-hand side of the equations in (3.1) are a combination of the Pb in MORB (mid-oceanic ridge basalt: this contains little crustal material, and is thought to be most suitable for investigating the composition of the upper mantle) and the Pb in lumps of manganese (known as Mn nodules) collected from the sea floor. Vollmer took the view that the Pb in MORB represents the Pb in the present upper mantle and the Pb in Mn nodules represents the average value of crustal Pb. He assumed that the mixture of the two corresponds to the primitive mantle remaining after the Fe core had just separated from the primitive earth (i.e., the mantle before separation of the earth's crust). Hence the t in the equations in (3.1) corresponds to the time at which the primitive mantle and the core separated from the homogeneous primeval earth. This argument contains some points that must be

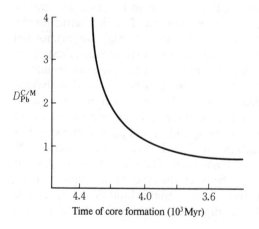

Fig. 3.2. Relation between a distribution coefficient ($D_{Pb}^{C/M}$) of Pb between metalic iron (core) and silicate (mantle) and age of the core separation (in millions of years). (After Vollmer, 1977)

resolved in the future, such as the appropriateness of applying the Pb distribution coefficient found in laboratory experiments to the interior of the earth and the validity of regarding the Pb in Mn nodules as the representative value for crustal Pb, so we cannot simply conclude that the value of t = 4200 million years deduced from Fig. 3.2 is the time at which the core was formed. Even taking these factors into account, however, it seems appropriate to conclude that, as far as we can tell from Pb isotopic ratios, formation of the core was completed at quite an early stage in the earth's evolution, more than 4000 million years ago at the latest. The fact that the existence of the earth's magnetic field can be traced back to 3800 million years ago, so that the Fe core that is the source of this geomagnetic field must have existed prior to that time, supports the conclusions derived from the Pb isotopic ratios.

3.6 Mantle Differentiation

As discussed in the previous section, we have considered that after it had just accreted from the well-mixed, homogenized solar nebula the primeval earth was virtually homogeneous from its surface down to its center. We have discussed the evolution of the earth based on the homogeneous accretion model. Eventually iron separated from the homogeneous primeval earth and formed a core in the center of the earth. From Pb isotopic data it was estimated that the formation of the core was completed in quite a short time after the birth of the earth, probably within several hundred million years. Immediately after separation of the core the earth consisted of a two-layered structure in which a metallic core lay at the center surrounded by a mantle consisting of silicates. The mantle at this time differed from the present mantle, as the earth's crust had not yet differentiated, i.e., it was a primeval mantle. Its chemical composition also differed considerably from that of the present mantle. When and how did this primeval mantle differentiate into the crust-mantle system of today? These questions will be the focus of discussion in this section.

Since one of the parties involved in the evolution of the mantle-crust system, i.e., the crust, lies within our reach, we can expect a more accurate discussion of the evolution of this system than in the case of the separation of the core. Let us first look at this issue from the earth's crust. As already discussed in Chapter 2, the average composition of the earth's crust is thought to resemble a mixture of basalt and granite in the ratio of approximately 3 : 1. We cannot always conclude that this granite is of mantle origin, but there is no doubt that basalt is the result of magma formed within the mantle erupting to the surface of the earth, where it solidified. Hence we can state that the history of the formation of the earth's crust amounts to an investigation of the manner in which basalt and granite were formed. Basalt was formed by mantle material partially melting to form magma, which then erupted to the surface of the earth and solidified. It is thought that most granite was also formed in the mantle and then brought to the surface of the earth. Let us consider the age distribution of these rocks. Granite is distributed throughout virtually all geological ages, from granite more than 3000 million years old produced in continental shields through to young granite formed in the Cenozoic era (< 65 million years). Volcanoes are susceptible to weathering and erosion after erupting to the surface of the earth, and almost no volcanoes older than a few hundred million years remain today. Nevertheless, volcanic rocks are distributed more or less continually throughout time from the present back through until the Precambian period. Viewed in this light, we can validly state that the earth's crust has been formed more or less continuously throughout all geological eras by means of a unilateral supply of material (magma) from the mantle to the crust. This viewpoint is known as the continuous crustal growth hypothesis.

In contrast to the continuous crustal growth hypothesis, there is another view that holds that the earth's crust was formed in the initial stages of the earth's evolution. In the past this view was based on the claim that granite was formed through the recrystallization of crustal material, and these claims were further reinforced by the concept of subduction introduced by plate tectonics. Magma brought to the surface from the mantle below mid-ocean ridges forms the ocean floor and spreads out on both sides. The spreading ocean floor eventually collides with the landmasses on both sides, and subsides beneath the continents. The areas where this occurs are known as subduction zones, and the Japan trench and the Izu/Mariana trench are typical examples. The ocean floor plate subsides into the trench, but deep sea sediment will also subside along with this. Material that has been carried to the ocean floor by erosion from the earth's surface accounts for a considerable proportion of this sediment. Hence the exchange of material between the mantle and the earth's crust must be regarded not merely as one-way traffic from the mantle, but as including a transfer of material from the earth's crust to the mantle, i.e., a literal exchange of material. The extreme view of this concept maintains that the earth's crust was formed in one go in the initial stages of the earth's history, and that the subsequent evolution of the crust-mantle system was no more than a recycling of material between the mantle and crust. This is known as the mantle-crust recycling hypothesis advocated by R. Armstrong and others.

Arguments over the evolution of the present mantle-crust system amount to a choice between the continuous evolution hypothesis and the recycling hypothesis, although an intermediate viewpoint also exists. Isotopic earth science methods, such as Pb, Sr, and Nd isotopic ratios, have made the largest contribution toward throwing light on the evolution of the mantle-crust system. Let us commence with a discussion of Nd isotopic ratios.

a) Nd Isotopic Ratios and Mantle Evolution

Together with Sr and Pb isotopic ratios, Nd isotopic ratios provide the most important lead to clarifying how the mantle evolved into its present form. It is only comparatively recently, however, that the use of Nd isotopic ratios has been introduced into earth science because of the experimental difficulties involved in carrying out isotopic analyses of Nd, which is a rare-earth element. The chemical properties of rare-earth elements are extremely similar. and so the 16 rare-earth elements are grouped together on the periodic table in one block. When analyzing isotopic ratios it is generally necessary to separate in advance other elements with the same mass number, but since rare-earth elements closely resemble each other chemically, it is extremely difficult to separate them from each other. The chemical affinity of rare-earth elements makes analyzing isotopic ratios difficult, but on the other hand this property is a great advantage when examining large-scale material differentiation, such as the evolution of the mantle. The reason for this will become clear later.

The Nd that exists in nature consists of seven isotopes with the mass numbers of 142, 143, 144, 145, 146, 147 and 148. Of these, ^{143}Nd contains radiogenic components formed through the α-disintegration of ^{147}Sm, another rare-earth element. In accordance with the method used by G.J. Wasserburg, let us use the ^{143}Nd/^{144}Nd isotopic ratio of meteorites as the criterion for discussing the evolution of the mantle, taking the stable isotope ^{144}Nd as the denominator and the radiogenic isotope ^{143}Nd as the numerator. We will indicate the present time as t = 0 on the time axis, and go back into the past and measure the time t. Generally speaking, the parent bodies of meteorites and meteorites themselves have existed as "frozen" bodies ever since their birth. Hence changes in the ^{143}Nd/^{144}Nd isotopic ratio of meteorites are determined unequivocally by the ^{147}Sm/^{144}Nd ratio of the meteorite at the time of its birth. By expressing the value at the time of the meteorite's birth by the subscript 0 and the present value by the subscript p, the value of (^{143}Nd/^{144}Nd) t years ago can be written as

$$(^{143}Nd/^{144}Nd)_t = (^{143}Nd/^{144}Nd)_p - (^{147}Sm/^{144}Nd)_p (e^{\lambda t} - 1).$$

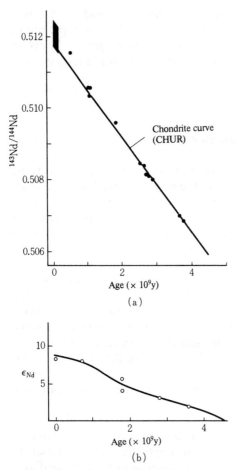

Fig. 3.3. a Evolution of $^{143}Nd/^{144}Nd$ isotopic ratio in chondrites (*solid line*) which also represents that in the mantle. *Solid circles* and a *lozenge* (upper left) indicate silicic crustal rocks (granite, granodiorites and related rock types), suggesting that the isotopic history of the upper mantle may be recorded in the silicic crustal rocks as well as in volcanic rocks. (After DePaolo, 1981). **b** Initial ε_{Nd} versus age for mantle-derived materials. Note that a systematic increase in ε_{Nd} with age. (After DePaolo, 1983)

Here λ is the decay constant of ^{147}Sm. If we consider the age of the earth $(4.5 \times 10^9 \text{ y})$, $\lambda t \ll 1$, and $e^{\lambda t} - 1 \cong \lambda t$, and the equation above can be written more simply as

$$(^{143}Nd/^{144}Nd)_t = (^{143}Nd/^{144}Nd)_p - (^{147}Sm/^{144}Nd)_p \cdot \lambda t. \qquad (3.3)$$

Figure 3.3a shows the line of growth of the Nd isotopic ratios in Eq. (3.3). As the figure reveals, the $(^{143}Nd/^{144}Nd)$ ratio in the meteorite has risen linearly from the value (known as the primordial value) at the

Mantle Differentiation

end of the nucleosynthesis approximately 4500 million years ago right up until the present.

This has been a general discussion, not supposing any special meteorite. As stated in Chapter 2, meteorites and planetary material can be regarded as having a more or less uniform relative elemental abundance, at least for nonvolatile elements. Sm and Nd are both nonvolatile elements and both belong to the rare-earth element group. They are little affected by chemical differentiation and their Sm/Nd ratio is virtually constant in any kind of meteorite. Hence the above argument and the relation shown in Eq. (3.3) can be regarded as generally valid regardless of the type of meteorite. On the basis of the meteorite analogy it also seems that there is no obstacle to regarding the Sm/Nd value in Eq. (3.3) as the average value for the earth. Hence the changes in the value of ^{143}Nd/^{144}Nd in Eq. (3.3) can be said to show the general Nd isotopic ratio growth common to all planetary material. The fact that the growth of ^{143}Nd/^{144}Nd is linear is simply a reflection of the fact that ever since their birth meteorites have been a closed system as far as Nd and Sm are concerned.

Next let us consider changes in the ^{143}Nd/^{144}Nd ratio in the mantle. An examination of volcanic and other rocks derived from the mantle will give us the ^{143}Nd/^{144}Nd ratio in the mantle. In order to learn about time variations in the isotopic ratio it is necessary to examine rocks of various ages. It must be noted here, however, that the older the rock is, the more in-situ decayed radiogenic ^{143}Nd it will contain as the result of the Sm in the rock. Hence the remainder, after having subtracted this radiogenic portion from the analyzed ^{143}Nd/^{144}Nd value, is the isotopic ratio of the mantle t years ago. Let us write the resulting isotopic ratio in the mantle as $I_M(t)$ ($= {}^{143}$Nd/^{144}Nd). For the sake of comparison, we will write the value found for meteorites in Eq. (3.3) (left-hand side of the equation) as $I_{met}(t)$. Owing to the long half-life of 10^{11} years for ^{147}Sm, even over such a long period as the history of the earth the change in the ^{143}Nd/^{144}Nd ratio is quite small. Let us therefore define the difference between the ^{143}Nd/^{144}Nd values of meteorites and the mantle, multiplied by 10 000, as

$$\varepsilon_{Nd}(t) \equiv \left\{\frac{I_M(t)}{I_{met}(t)} - 1\right\} \times 10^4, \tag{3.4}$$

and use this value in the following discussion. Figure 3.3b shows the ε_{Nd} values found so far for mantle materials of various ages. Let us now consider the evolution of the mantle using Fig. 3.3b.

For meteorites, by definition, $\varepsilon_{Nd}(t) = 0$, and this is the t-axis in Fig. 3.3b. As time passes, the data for the mantle gradually shift upwards from the t-axis. The fact that the value of ε_{Nd} shifts in the plus direction with time shows that the increase in the ^{143}Nd/^{144}Nd ratio is more marked in the mantle than in meteorites, and that hence the ^{147}Sm/^{144}Nd value in the mantle is larger than in meteorites. What process in mantle evolution lies behind this?

Here we will consider this question from the viewpoint of the continuous evolution model, one of the representative viewpoints in mantle evolution theory. According to this hypothesis, the earth's crust has been formed by receiving a continuous supply of material from the mantle throughout geological time. Let us suppose that mantle material melted partially and formed magma. Trace elements (elements that have a low abundance and cannot form crystals on their own) such as Nd and Sm would be selectively incorporated in the melt, and so they would be relatively deficient in the crystals remaining after this partial melting compared to before the melting, that is, we have element fractionation. Both Nd and Sm are trace elements (mantle material contains only about several ppm at the most), and at the time of partial melting they are overwhelmingly enriched in the melt phase, but Nd is enriched even more selectively than Sm. Consequently, as partial melting (in order to form basalt magma it is necessary to melt approximately 1–20% of the mantle material) of the mantle proceeded, the solid phase of the mantle gradually would become deficient in Nd and Sm, and the deficiency of Nd would become more marked than that of Sm. The issue of the manner in which trace elements are distributed between the solid phase (crystals) and the liquid phase (magma) at the time of partial melting is known as trace element distribution theory, and is an extremely important area of basic research in geochemistry. A detailed discussion of this lies beyond the scope of this book, so the reader should consuet the references (e.g., Broecker and Oversby 1971) for further information.

Let us consider the state of the earth immediately after the core had separated from the homogeneous earth. The metal core consisting mainly of Fe contains almost no Nd or Sm. Hence the Sm/Nd ratio in the primitive mantle remaining after the core had separated has virtually the same value as meteorites and as the earth at the time of its birth. Take a look at Fig. 3.3b. When the primitive mantle was first formed it had a ^{147}Sm/^{144}Nd ratio more or less the same as that of meteorites. Hence the initial ε_{Nd} of the mantle will appear on the t-axis in the figure. As time passes, the repeated partial melting of the mantle will lead to a gradual reduction in the amounts of Sm and Nd in the mantle. Since the decrease of Nd will be greater than that of Sm, however, in the long run the Sm/Nd ratio in the mantle will rise gradually over its previous value. Corresponding to this rise, we can expect the ^{143}Nd/^{144}Nd ratio also to rise as time passes. Thus the upward tendency of the left-hand side in Fig. 3.3b can be explained if we consider how the process of partial melting, magma formation, eruption of volcanic rocks and crustal formation has been repeated throughout geological time to produce the earth's crust. Further examination reveals, however, that the rate of increase in $\varepsilon_{Nd}(t)$ is considerably smaller than the rate of increase expected from actual estimates of the Sm/Nd ratio of the mantle. This discrepancy indicates that when considering the evolution of the mantle we must take into account not only the unilateral movement of material from the mantle to the earth's crust, but also the return flow of material from the earth's crust to the mantle. From the data in Fig. 3.3b, D.J. De Paolo has estimated the amount of this material movement from the earth's crust to the mantle as $(0.35 \pm 0.15) \times M_c/10^9$ y (M_c is the mass of the present earth's crust). This is equivalent to the whole of the earth's crust completely sinking into the mantle over a period of approximately 3000 million years.

b) Sr Isotopic Ratio in the Mantle

So far we have discussed the development of the mantle-crust system based on the $^{143}Nd/^{144}Nd$ ratio. A similar discussion can be developed using the Sr isotopic ratio of $^{87}Sr/^{86}Sr$, where the isotopic ratio changes with time due to $^{87}Rb \rightarrow {}^{87}Sr$. We can replace ^{147}Sm and ^{143}Nd by ^{87}Rb and ^{87}Sr respectively and select ^{86}Sr as the stable isotope instead of ^{144}Nd. We can then develop the argument in the same manner as in the case of Nd, based on the Sr isotopic ratio of $I(t) = {}^{87}Sr/^{86}Sr$. However, the Rb–Sr system differs from the Sm–Nd system in that Rb is a volatile element. As has been stated repeatedly, we can assume that meteorites and planetary material have roughly the same relative abundances, at least as far as nonvolatile elements are concerned. In the case of Rb and other volatile elements, however, this meteorite analogy cannot be applied as is. Rather, various supporting evidence suggests that not only did the Rb/Sr ratios of meteorites and the earth differ to a considerable extent, but also that the ratio differed amongst meteorites. Hence, we can neither define an average Rb/Sr ratio for meteorites nor suppose that the average value, if defined, is the same as that in the earth.

In contrast to Sm and Nd, Rb and Sr are quite different chemically, and they would have acted quite differently when the mantle evolved. In the case of partial melting of the mantle, for instance, the Rb/Sr ratios in the magma and the solid phase would differ greatly. From laboratory experiments it is anticipated that the Rb/Sr ratio of magma would be more than ten times that of the solid phase. This is in sharp contrast to the Sm/Nd ratio, which changed only a fraction in response to partial melting of the mantle. There is also a major difference between the $^{87}Sr/^{86}Sr$ ratio of the mantle and that of crustal material. Reflecting the large Rb/Sr value, it is normal for crustal rocks to have a higher $^{87}Sr/^{86}Sr$ ratio than mantle material.

Because of the chemical similarity of Sm and Nd, the ^{143}Nd/^{144}Nd ratio may be suited to dealing with large-scale phenomena such as the separation of the mantle and the earth's crust, whereas the ^{87}Sr/^{86}Sr ratio is more responsive to phenomena such as the mixing of crustal material and mantle material. Thus combining these two ratios provides even more valuable data on the evolution of the earth. Let us therefore define ε_{Sr} similarly to the definition of ε_{Nd}. In the case of Sr, however, we are unable to define a common meteorite value of $I_{met}^{Sr}(t)$ as stated above, so instead of using meteorites we will follow Wasserburg by defining a hypothetical material UR (uniform reservoir) that has a ^{87}Sr/^{86}Sr value of 0.7045 and a ^{87}Rb/^{86}Sr value of 0.0839. The (^{87}Sr/^{86}Sr)$_{UR}$ value was determined as the one corresponding to $\varepsilon_{Nd} = 0$ in a ε_{Nd}–ε_{Sr} correlation diagram (Fig. 3.4). The (^{87}Rb/^{86}Sr)$_{UR}$ value was then estimated as the one which brings the initial ratio (^{87}Sr/^{86}Sr)$_0$ to the (^{87}Sr/^{86}Sr)$_{UR}$ in 4.5 billion years. Therefore ε_{Sr} is defined similarly to ε_{Nd} as

$$\varepsilon_{Sr} = \left(\frac{I_M^{Sr}(t)}{I_{UR}^{Sr}(t)} - 1 \right) \times 10^4.$$

Here I_{Sr} is the ^{87}Sr/^{86}Sr ratio, and M and UR express the values in the mantle and the uniform reservoir. Using these definitions of ε_{Nd} and ε_{Sr}, let us consider a $\varepsilon_{Nd} - \varepsilon_{Sr}$ figure in which ε_{Nd} is plotted on the vertical axis and ε_{Sr} on the horizontal axis. On the coordinates we will plot the ε_{Nd} and ε_{Sr} data found for very young volcanic rocks, i.e., data corresponding to $\varepsilon_{Nd}(0)$ and $\varepsilon_{Sr}(0)$. The results that have been obtained so far are compiled in Fig. 3.4. As the figure reveals, the data for the volcanic rocks lie on a virtually straight line, but the lower right-hand part diverges from this linear array. Before moving on to an interpretation of these volcanic rock data, let us examine closely the significance of the $\varepsilon_{Nd} - \varepsilon_{Sr}$ plot.

As the origin of the coordinate axes in the $\varepsilon_{Nd} - \varepsilon_{Sr}$ plot, we chose for ε_{Nd} a value that uses meteorites as the reference [by definition $\varepsilon_{Nd}(t) = 0$ for $t \geq 0$] and for ε_{Sr} a value that uses a UR as the reference. This choice was made so that the origin of the coordinate axes would correspond roughly to the average composition of the earth. The meteorite analogy means that this hypothesis can be regarded as perfectly appropriate at least for ε_{Nd}, but there is no guarantee as regards ε_{Sr}. Let us suppose for now that $\varepsilon_{Nd} = 0$ and $\varepsilon_{Sr} = 0$ correspond to the average values for the whole earth. If we make this assumption, how are the isotopic data in Fig. 3.4 to be interpreted? Let us consider this first from the viewpoint of the continuous crustal growth model, i.e., the process by which the homogeneous primeval mantle formed magma and the earth's crust differentiated gradually. As stated earlier, in magma the Rb/Sr ratio increases and the Sm/Nd ratio decreases. In the mantle remaining after the magma has been extracted, conversely the Rb/Sr ratio will decrease

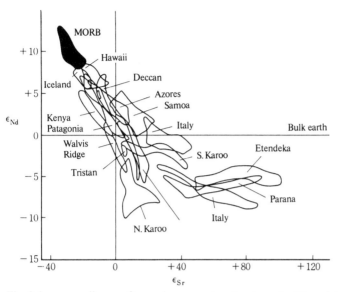

Fig. 3.4. $\varepsilon_{Nd}-\varepsilon_{Sr}$ diagram for modern oceanic volcanic rocks. The origin corresponds to a hypothetical bulk composition of the earth. (After Hawkesworth et al., 1984)

and the Sm/Nd ratio will rise. As a result ε_{Nd} will increase and ε_{Sr} will decrease in the mantle. In effect, the MORB in the upper left-hand side of Fig. 3.4 can be regarded as the solid phase remaining after the magma has been extracted. This conclusion is the same as that reached in the previous section from ε_{Nd} alone. In Fig. 3.4, however, a large number of volcanic rocks are also distributed in the lower right-hand side of the $\varepsilon_{Nd} - \varepsilon_{Sr}$ plot, i.e., in the fourth quadrant. Volcanic rocks from Italy are a typical example. Data on basalt from the Deccan Plateau in India and on volcanic rocks from such places as Samoa and the Kerguélen Islands are arranged more or less linearly, as if to fill in the space between MORB and the Italian volcanic rocks. It is difficult to explain this tendency by a simple process consisting of a primeval mantle, magma eruption, mantle differentiation, and formation of the earth's crust.

Ever since its discovery, the linear array of ε_{Nd} and ε_{Sr} data shown in Fig. 3.4 has been referred to as mantle array, and has been regarded as a reflection of a fundamental process in the evolution of the mantle. In explaining this mantle array, it is normal nowadays to consider it as a mixed line between the differentiated MORB that corresponds to the mantle and the components on the lower right-hand end. Various theories have been put forward as to the components on the lower right-hand end of the mantle array – fragments of oceanic crustal material that have sunk into the mantle through subduction, or continental lithosphere that has sunk into the mantle as the result of mantle convection, or simply material formed through a mixture of crustal material and mantle material in the process of magma rising up to the earth's crust. No definite conclusion has been reached, however, about the nature of this material. Nevertheless, the $\varepsilon_{Nd} - \varepsilon_{Sr}$ mantle array does demonstrate at least that the evolution of the mantle cannot be explained by a simple continuous evolution model consisting of a one-way transfer of material from the mantle to the earth's crust. We must also take into account the return to the mantle of a certain amount of material from the earth's crust.

3.7 The Age of the Mantle

In our discussion so far we have assumed that the newly accreted earth had a roughly homogeneous composition. We have considered that shortly after its birth ($\lesssim 10^8$ years) the earth differentiated into core and primitive mantle, and that the earth's crust separated from the primitive mantle and has been formed more or less continuously throughout the whole evolution of the earth. So far, however, we are not sure whether the formation of the earth's crust involved the whole mantle or merely one region of the mantle. If the earth's crust differentiated and formed from only one part of the mantle, the possibility exists that an undifferentiated region remains within the mantle. Does this mean that the undifferentiated mantle that did not contribute to the formation of the earth's crust would have an extremely old age that reflects the time of separation of the core and the primitive mantle? Even apart from this issue, the question of the age of the rocks in the depths of the earth is of sufficient interest by itself. This section will discuss the age of mantle rocks.

Radiometric dating is the most direct method of determining the age of the mantle if representative mantle materials can be obtained. However, the radiometric age of material that did not become a closed system with regard to radiogenic elements until the material erupted to the surface of the earth merely gives us the date at which the rocks erupted, even though the material is derived from the mantle. In order to ascertain the age of the mantle, the sample used must be material that was formed in the mantle and has existed ever since then as a closed system. Mantle xenoliths could be such material. Attempts have been made to date numerous mantle xenoliths using such methods as the K–Ar, Rb–Sr, and Sm–Nd methods. Nearly all of these attempts, however, have met with failure. The sole reason for this is the fact that the mantle is in a state of high temperature ($> 1000\,°C$), and this has accelerated the diffusion of elements, and so the prerequisite of a closed system has not

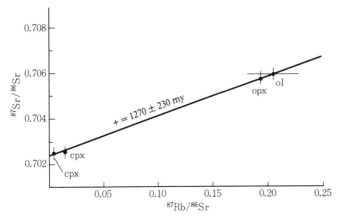

Fig. 3.5. Rb–Sr isochron plot for minerals separated from a New Mexico ultramafic xenolith. *opx* orthopyroxene; *cpx* crynopyroxene; *ol* olivine. (After Steuber and Ikramuddin, 1974)

been fulfilled. The only exception so far appears to be the successful Rb–Sr dating by A.M. Steuber and M. Ikramuddin of a mantle xenolith contained in basalt. The results of this dating are shown in Fig. 3.5. The sample was an ultrabasic xenolith included in basalt found in New Mexico. Olivine, orthopyroxene, and clinopyroxene were separated from the xenolith, and their Rb–Sr mineral ages were found by the isochron method. Each sample lines up neatly on a straight line on the $^{87}Sr/^{86}Sr - ^{87}Rb/^{86}Sr$ isochron plot. A value of t = 12.7 ± 2.3 hundred million years was found from the slope of the straight line. The basalt containing this xenolith is volcanic rock that erupted within the last several million years. The figure of 1270 million years may correspond to the age at which this xenolith crystallized within the mantle and began to exist as a closed system.

Improved mass spectrometry techniques have led to attempts in recent years to date diamonds, which are the most symbolic sample of mantle origin available to us.

In most cases diamonds have some fine inclusions of such distinctive minerals as olivine and garnet. Since these inclusions contain some Sm, Rb, and K, they can be dated radiometrically. Rb–Sr, Sm–Nd, and Pb–Pb dating of the inclusion materials, such as the olivine and garnet that have separated from diamonds, has been carried out in the past. The amount of the inclusions in diamonds and also the amount of such elements as Rb and Sm in the inclusions are very small, making it difficult to perform experiments, so there has been no successful dating using the isochron method so far. However, there are a few cases in which a "model age" has been found by assuming reasonable primordial

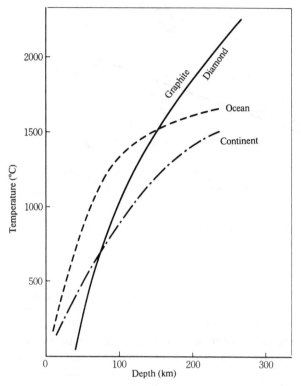

Fig. 3.6. Equilibrium diagram (*solid curve*) for a graphite-diamond system. Also shown are geotherms for oceanic mantle (*dotted*) and continental mantle (*dash-dotted*)

values for the isotopic ratios of Sr and Nd. Using the Sm–Nd system, recently S.H. Richardson and his colleagues have reported a model age of approximately 3000 million years for the inclusions in diamonds of South African origin. From the configuration and composition of these inclusions, Richardson et al. concluded that they were formed at roughly the same time as the diamonds crystallized.

The diamonds used by Richardson and his colleagues for dating were contained in kimberlite that is estimated to have erupted about 90 million years ago, which is a far younger age than that found for the diamonds themselves. Diamonds are a high-pressure form of carbon crystal produced in the depths (a few hundred kilometers) of the earth. Figure 3.6 shows the phase diagram of diamonds, as well as the temperature distribution curve for the earth's interior. As the figure reveals, diamonds are stable at depths below about 150 km, but at shallower sites they become unstable and undergo a phase transition to graphite. Moreover, it is shown that this phase transition occurs very rapidly under the high-temperature conditions in the earth's interior. Therefore, if diamonds that have crystallized in the depths of the earth were brought up to a depth of less than 100 km as the result of mantle convection, they would break down and become graphite. Although the diamonds examined by Richardson and his colleagues erupted to the surface of the earth less than 90 million years ago, they actually crystallized more than 3000 million years. Hence these diamonds existed in the mantle for almost 3000 million years ago, indicating that the diamonds and the mantle region where they crystallized never moved to the upper mantle above a depth of 100 km throughout this period of approximately 3000 million years. This conclusion imposes an important constraint on mantle evolution, particularly on mantle convection.

It has frequently been claimed that corresponding to the layered structure of the mantle concluded from seismic waves, mantle convection also occurs independently in the upper and lower parts of the mantle. The results on the diamonds by Richardson and his colleagues may be relevant to these claims.

Recently Ozima and Zashu have measured the isotopic ratios of the He contained in diamonds, and the results have led us to suppose that a certain kind of diamond may be roughly as old as the earth. Diamonds contain $10^{-6} - 10^{-9}$ cm^3 STP g^{-1} of He. The isotopic ratio ^{3}He/^{4}He is distributed more or less continuously from 10^{-7} up to 3.4×10^{-4}. Since the formation of ^{3}He under the conditions on earth can be virtually disregarded, most ^{3}He can be regarded as being of primordial origin, having been captured at the time of the earth's formation. Currently, two basic He components are known to exist in the solar system. One, known as solar He, is observed in the solar wind, and has an isotopic ratio of ^{3}He/^{4}He $= 4 \times 10^{-4}$. The other He component is found characteristically in meteorites, and has a value of ^{3}He/^{4}He $= 1.4 \times 10^{-4}$. The latter is the He that was formed in the nucleosynthesis, and is thought to be the most primitive He in the solar system. This He is customarily referred to as planetary He, but it has no direct connection with the planets. The higher ^{3}He/^{4}He ratio of solar He compared to that of planetary He is attributable to the deuterium burning (^{2}D + ^{1}H $\rightarrow$ ^{3}He + γ) that occurred in the sun immediately after its formation ($< 10^7$ years). At this stage no conclusion can be drawn as to which type of He was captured when the earth was formed, although the results of research into diamonds suggest that the earth captured a considerable amount of solar He, but in any case it is certain that the ^{3}He/^{4}He ratio at the time of the earth's formation was $> 10^{-4}$.

As discussed in Section 3.2, the earth is supposed to contain approximately 20 ppb of U and 80 ppb of Th. Hence ^{4}He produced through the disintegration of U and Th is accumulated in the earth's interior, and as time passes the ^{3}He/^{4}He isotopic ratio in the earth's interior will gradually decrease. The manner of this decrease depends not only on the amounts of U and Th in the earth, but also on the value of the ^{3}He initially incorporated in the earth. Assuming a value of 20 ppb for U and of 10^{-11} cm^3 STP g^{-1} for ^{3}He, from the heat flow on the surface of the earth and from the meteorite analogy and calculations of the change in ^{3}He/^{4}He, it can be demonstrated that within several hundred million years the value of ^{3}He/^{4}He in the interior of the earth will fall considerably below 10^{-4}. In most cases U and Th in diamonds are virtually negligible, and there would be almost no change in the ^{3}He/^{4}He isotopic ratio once He has been trapped within a diamond – the ^{3}He/^{4}He isotopic ratio is "frozen". Consequently, the ^{3}He/^{4}He isotopic ratio of diamonds which contain an extremely small amount of U and Th can be regarded as a faithful reflection of the value in the mantle when the diamond crystallized. Hence the fact that some diamonds (so far four diamonds with a ^{3}He/^{4}He ratio of more than 2×10^{-4} have been detected, all from South Africa) have a ^{3}He/^{4}He isotopic ratio close to the primitive value would signify that the He was captured within these diamonds before the ^{3}He/^{4}He isotopic ratio changed significantly in the earth's interior, probably within several hundred million years of the earth's formation. This reasoning suggests that some diamonds crystallized not long after the formation of the earth.

Dating of xenoliths and diamonds derived from the mantle suggests that the mantle regions in which these samples crystallized have existed stably over thousands of millions of years. Currently we have no clear knowledge of the extent of these old and stable regions in the present mantle. Some researchers are of the opinion that they are distributed in the mantle in dispersed patches, but many other scientists believe that only the upper mantle contributed to the formation of the earth's crust, and that the lower mantle remains undifferentiated. Convincing proof for this is the fact that the primordial He isotopic ratio (^{3}He/^{4}He > 1×10^{-5}) is characteristically found at hot spots, which are generally thought to build oceanic volcanoes by supplying magma directly from the lower parts of the mantle. Following this line of thought, it means that the lower part of the mantle retains its original state. This will be discussed in further detail in Section 3.9.

3.8 Origin and Evolution of the Atmosphere and Oceans

As will become gradually clear in this section, both the earth's atmosphere and the oceans were derived secondarily from the earth's interior after the birth of the earth. It is expedient, therefore, to treat the atmosphere and oceans as a single entity from a geohistorical point of view and to discuss them in comparison with the solid earth. Here we will collectively label as the "atmosphere" the air (the atmosphere in a narrow sense), oceans (including rivers and lakes), sedimentary rocks and volatile components such as H, C, N, S, Cl and rare gases that are included in the biosphere.

a) Secondary Origin of the Atmosphere

Figure 3.7 takes up H, C, N, S, Cl, and rare gases as the main volatile components of the atmosphere, and plots the relative concentrations of these elements with regards to their solar abundances. As is obvious from this figure, the atmospheric abundances of all of these volatile components are less than their solar abundances by two (Cl) to more than ten orders of magnitude (Ne). Let us suppose for now that a part of the solar nebula was trapped in the earth's gravity field and remained as the earth's atmosphere. The composition of the resulting atmosphere would resemble that of the solar nebula. As Fig. 3.7 shows, however, there is virtually no correlation between these compositions. To explain the relative deficiency of volatile components in the earth's atmosphere compared to their solar abundances, we may argue that some volatile components remain inside the solid earth. The stronger the coherence with solid earth material, i.e., the stronger is the element's chemical affinity, the easier it should be for these components to remain inside the

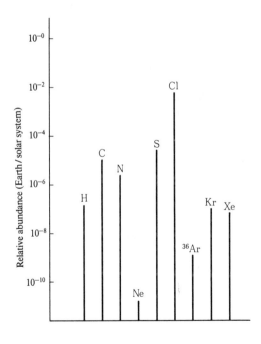

Fig. 3.7. Ratios of observed abundances of atmospheric (volatile) elements to abundances expected for cosmic composition. Note an enormous depletion of noble gases relative to other volatiles. (After Ozima and Podosek, 1983)

earth. Completely contrary to these expectations, however, the results of Fig. 3.7 show that rare gases that have no chemical affinity, i.e., ones which have more difficulty in remaining in the solid earth, are far more deficient in the atmosphere than volatile components that have chemical affinity. Conversely, it is also difficult to attribute the relative deficiency of the volatile elements to their escape from the atmosphere into the outer space. In the latter case, we expect that the lighter volatile elements would escape more efficiently than the heavier ones, which is contrary to oberservation (Fig. 3.7). It would be no exaggeration to say that this indicates more or less conclusively that the origin of the earth's atmosphere cannot be sought in the solar nebular gas. This view was first expressed by H. Brown in 1950, and is the starting point in considering the origin of the earth's atmosphere. This was the basis for the theory of the secondary origin of the earth's atmosphere.

If the present atmosphere is not a primary atmosphere that was left over from the solar nebula, then where, when and how was it formed? In a classic paper entitled *Geological History of Sea Water*, which was published at about the same time as the classic paper by Brown, W.W. Rubey answered these questions as follows. Rubey first focused his attention on the fact that the composition of the earth's atmosphere resembles that of volcanic gas and fumarolic gas. He pointed out that if volcanic gas, hot spring waters and fumarolic gas have been released throughout geological time at about the same ratio as today, the total amount would easily exceed the total volume of the earth's atmosphere. He therefore concluded that the atmosphere has been drawn from the mantle throughout geological time as the result of volcanic and other activity. Rubey's theory of the origin of the atmosphere is known as the continuous degassing model, and it supposes that the atmosphere has been formed by degassing from the earth's interior throughout geological time. At one time the continuous degassing model was the principal theory on the origin of the earth's atmosphere.

Subsequently, however, it has been pointed out that many details of the continuous degassing model are incompatible with actual observations, especially the results of geochronology. For example, the results of Rb–Sr dating carried out by S. Moorbath and the Oxford group from the late 1960's onwards have revealed that the metamorphic rocks that are widely distributed on the west coast of Greenland are about 3800 million years old. The source rocks of these metamorphic rocks are obviously sedimentary rocks, indicating that more than 3800 million years ago oceans already existed to a considerable extent. Moreover, the composition of sedimentary rocks 2000 to 3000 million years old, including these metamorphic rocks in Greenland, does not differ greatly from the composition of present sedimentary rocks, suggesting that the sedimentation environment at that time, i.e., the properties of the atmosphere and oceans, did not differ greatly from that nowadays. Three thousand million years ago the earth had lived through only one-third of its history, and according to Rubey's continuous degassing model only one-third of the atmosphere should have been formed. This is clearly incompatible with the above circumstantial proof based on earth science considerations. On the basis of several such pieces of geological circumstantial proof, F.P. Fanale claimed that the formation of the atmosphere occurred catastrophically at a very early stage in the earth's evolution and over an extremely short period.

The isotopic ratios of rare gases, particularly the Ar isotopic ratio, provide us with a quantitative insight into the mechanism behind atmospheric degassing. Before moving on to a discussion of the degassing of rare gases, let us review the state of the earth before the appearance of the secondary atmosphere. As stated in Chapter 2, the planets were formed by condensing from the solar nebula, and the birth of the earth cannot be separated from this solar nebula. In order to conclude that the earth's atmosphere has a secondary origin, we must demonstrate that the solar nebula that must have enveloped the earth at the time of its birth had more or less dissipated completely before the secondary atmosphere appeared. It is customary to assume that the solar nebula was dissipated by the solar wind during a period of extremely violent solar activity (known as the T Tauri stage) that is thought to have occurred in the very early stages of solar evolution. Astrophysical theory concludes that the T Tauri stage occurred approximately 10^7 years after the formation of the sun and at that time the sun emitted a very strong solar wind (about 10^4 times the strength of the current solar wind), dissipating the solar nebula that had remained in solar space until then.

As its name indicates, the T Tauri stage is an actual phenomenon observed on T Tauri stars. Newly formed T Tauri stars emit a large quantity of gas into space. Theoretical calculations also confirm that in the process of stellar evolution stars that have reached this T Tauri stage release a large quantity of gas in this violent manner.

The time at which the sun reached the T Tauri stage is a decisive factor governing the formation of the earth and its subsequent evolution. Adopting the scenario drawn up by C. Hayashi and his colleagues, it is concluded that under the existence of the solar nebula the growth from planetesimals to primitive planet proceeded quite rapidly (over a period of about 10^7 years). This is understood to be the result of the fact that the solar nebula acted as a resisting force to the movement of the planetesimals, thus reducing their velocity and enabling the collision and accretion of planetesimals to proceed more effectively. According to the calculations of Hayashi and his colleagues, the formation of the earth was completed in about 10^7 years, before the sun reached the T Tauri stage. They concluded that eventually the gravitational force of the earth, which had grown within the solar nebula, attracted a primary atmosphere of several hundred atmospheres. Wrapped in this thick primary atmosphere, the earth must have reached quite a high temperature immediately after its birth, preserving the energy produced through the collisions of planetesimals, i.e., the earth accretion energy, without releasing it outside. Chapter 2 discussed how this led to the formation of the earth's core in the very early stages of the earth's evolution.

In contrast to the Hayashi group, which claims that the planets formed in the presence of the solar nebula, the generally accepted theory of the formation of the solar system, represented by V.S. Safronov, argues that the planets formed within a vacuum after the solar nebula had been completely dissipated. Since gas resistance would not decelerate the velocity of the planetesimals if the planets formed in a vacuum, it would have taken more than ten times as long for the planets to form than if they had formed in the presence of the solar nebular gas. Hence planet formation would have occurred long after the T Tauri stage, and we are forced to discuss planet formation within a vacuum. Note, however, that this is by no means proof that the formation of planets occurred within a vacuum. We must not forget that the "accretion in a vacuum" theory typified by Safronov is a scenario drawn up to fulfil the requirements imposed by the observed facts that the earth's atmosphere is of secondary origin and that the primary atmosphere has been completely dissipated. The logical consequence of planetary formation within a vacuum will run into the fundamental difficulty that even in the vicinity of the earth a formation interval of over 10^8 years would be necessary. An interval of almost 10^{10} years would be necessary for regions around Uranus and Pluto, which are even more distant from the sun and where the solar nebular gas is sparse, and so far no such stars have been born.

The planets were formed from the solar nebula. In an environment in which solar nebular gas existed, the formation of the planets would have proceeded rapidly, being completed in about 10^7 years in the vicinity of the earth. This is very likely before the sun entered the T Tauri stage, and the scenario of planet formation in the presence of the solar nebula completes itself. The question of whether or not formation of the planets occurred under the existence of the solar nebula, i.e., whether or not a primary atmosphere existed on earth, is one of the most basic issues in geohistory and earth science. At present we have no evidence of the existence of a primary atmosphere. On the other hand, nor do we have any definite facts refuting its existence. All that can be said with any conviction about the origin of the earth's atmosphere is that the present atmosphere was drawn from the earth's interior and is not a remnant of the solar nebula. Settling the issue of the existence of a primary atmosphere must await the results of further research into this problem, which will be a central topic in the future study of early geohistory. Here we will proceed with our discussion based on the scenario of the Kyoto group, headed by C. Hayashi, which seems to be superior as far as logical consistency is concerned.

b) Degassing Model

From the above discussion we have learnt that the atmosphere present on earth today has been formed through degassing from the earth's interior after the primary atmosphere that was left over from the solar nebula had been dissipated. Then when and how did this degassing of the earth's atmosphere occur? The isotopic compositions of He, Ar, and Xe provide us with our most powerful lead in quantitative discussions of degassing. Here we will use the example of the Ar isotopic ratio to discuss the degassing model.

Ar accounts for approximately one percent of the present atmosphere, and is the third most common element in the atmosphere after N_2 and O_2. ^{40}Ar accounts for 99.6% of all the argon in the earth's atmosphere, while ^{38}Ar (0.0632%) and ^{36}Ar (0.3364%) make up the remainder. Other rare gases are highly deficient in the earth's atmosphere compared to their solar abundances, so what accounts for this abundance of Ar? C.F. von Weizsäcker, Nobel Prize physicist, reached the conclusion that the ^{40}Ar in the earth's atmosphere is the result of the ^{40}K in the solid earth disintegrating into ^{40}Ar through electron capture nuclear disintegration, and that this ^{40}Ar degassed from inside the earth and was drawn into the atmosphere. This was proposed in 1937. This paper could be regarded as a precursor of the degassing theory, but its significance for earth science failed to gain the attention of earth scientists of the time, as it was obscured by its interest from a physics point of view as an actual example of electron capture nuclear disintegration. The phenomenon of electron capture nuclear disintegration had been first predicted by H. Yukawa and S. Sakata in 1935, and was discovered by L.W. Alvarez 2 years later, and these results were at the forefront of nuclear physics at that time.

Currently the Ar in the earth's atmosphere has a value of ^{40}Ar/^{36}Ar = 295.5. From nuclear synthesis theory the ^{40}Ar/^{36}Ar value when the solar system was formed is estimated to have been approximately 10^{-4}. A value of ^{40}Ar/^{36}Ar < 10^{-2} has been reported for meteorite components containing an extremely small amount of K and for the solar wind. The discrepancy with the ^{40}Ar/^{36}Ar ratio observed for the Ar in the atmosphere today can be attributed to the disintegration of ^{40}K in the earth's interior. However, in order to understand why the ^{40}Ar/^{36}Ar ratio in the earth's atmosphere came to have the value of

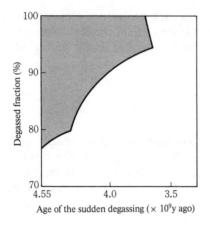

Fig. 3.8. Relation between a degassed fraction (%) and an age of degassing (millions of years). If we assume (i) 80 ppm < K < 400 ppm and (ii) ^{40}Ar/^{36}Ar > 5,000 in the present mantle, only the shaded area is allowed. (After Hamano and Ozima, 1978)

295.5, we must know the amounts of K and Ar in the earth, and when and how Ar degassed from the earth's interior into the atmosphere. For instance, if we suppose that K was contained in the early earth in relatively larger amounts than Ar, then the ^{40}Ar/^{36}Ar ratio can be expected to have a larger value. If the degassing from the earth's interior occurred at an extremely early stage in the earth's evolution, not much radiogenic ^{40}Ar would have been stored, and so the degassed Ar would also have had a small ^{40}Ar/^{36}Ar ratio. Conversely, this suggests that the characteristic isotopic ratio of the Ar in the present atmosphere provides an important clue to atmospheric degassing.

The isotopic ratio of the Ar in the atmosphere is not sufficient to impose definite constraints on a clarification of the degassing process. Y. Hamano and M. Ozima have shown that in addition to the isotopic ratio of the atmospheric Ar (^{40}Ar/^{36}Ar = 295.5), if the average ^{40}Ar/^{36}Ar ratio in the present mantle is known quite severe constraints can be imposed on the degassing of Ar from the mantle. Figure 3.8 shows these results. Here we are assuming that the degassing of Ar from the mantle occurred through two main processes, (i) a catastrophic degassing in the initial stages of the earth's evolution (t_d), and (ii) the continuous degassing that has occurred subsequently throughout the earth's evolution. Specific events corresponding to hypothesis (i) are the disturbance within the earth accompanying the core formation that is assumed to have occurred immediately after the formation of the earth, or the impact degassing when planetesimals collided with the earth in the later stages of the earth's formation. One possibility for hypothesis (ii) is the degassing accompanying the introduction of magma from the mantle to the earth's crust, which has occurred throughout the earth's evolution. In Fig. 3.8 the time (t_d) at which the catastrophic degassing occurred is shown on the horizontal axis, and the vertical axis shows the fraction of ^{36}Ar that was released into the atmosphere at that time from the earth's interior relative to the ^{36}Ar contained in the present atmosphere. If we allow the following two assumptions, (1) when Ar and K move with the magma from the mantle to the earth's crust and atmosphere, Ar would be transported more effectively than K, and (2) the K content in the mantle at present is 80 ppm $<$ K $<$ 400 ppm, and ^{40}Ar/^{36}Ar $>$ 5000 – then the portion marked by the shaded area in Fig. 3.8 is the only permissible region. This means that the degassing of Ar from the earth's interior occurred at an extremely early period ($t_d >$ 4000 million years ago), and that at least 75% of the Ar degassed at that time.

The ^{40}Ar/^{36}Ar ratio in the mantle can be estimated from the analysis of Ar trapped in glassy rims of submarine volcanic rocks. The gas in volcanic glass that has cooled rapidly on the ocean floor is trapped inside, unable to escape. Therefore volcanic glass from the ocean floor is extremely useful in examining gases of mantle origin. The latest results of analyses of the rare gases in this volcanic glass suggest strongly that the value in the present mantle (which was the reservoir for the atmosphere) has ^{40}Ar/^{36}Ar $>$ 10 000. In this case t_d will shift back to an even earlier time in the initial stages of the earth's evolution.

c) The Primary Atmosphere

Here our discussion of atmospheric degassing has focused on the isotopic ratios of rare gases. However, rare gases are trace components of the atmosphere, and such major components as N_2 and O_2 cannot be overlooked in a discussion of atmospheric evolution. The Ar degassing theory suggests that most of the atmospheric Ar was degassed catastrophically in the early stages of the earth's evolution (although ^{40}Ar produced through radioactive decay has been formed throughout the whole history of the earth, and so it has also degassed continuously throughout the earth's evolution). Naturally, this violent degassing would have involved other volatile components also. We may regard the earth's atmosphere as having been degassed more or less completely in the early stages of the earth's evolution. Then what was the composition of the primitive atmosphere that was more or less completely formed in the early stages of the earth's evolution?

The main components of the present atmosphere are N_2 and O_2. Researchers are in general agreement that O_2 was produced through green plant photosynthesis. It is generally thought that there was no O_2 in the primitive atmosphere that existed before life appeared on earth. Opinion is divided, however, on the question of whether or not N_2 was present in the primitive atmosphere. At one time there was an influential theory that the primary atmosphere of the earth had an extremely reducing composition consisting mainly of CH_4 and NH_3. This reducing atmosphere was a conclusion that suited biologists very well, dealing as they do with the origin of life, since the first step in producing life requires that organic matter be synthesized from inorganic matter. In the 1950's S.L. Miller succeeded in synthesizing amino acids, which are indispensable to the origin of life, from inorganic matter. The inorganic matter used in this experiment was a reducing gas consisting of a compound of CH_4, NH_3, H_2O, and H_2. Miller discharged sparks in this gas and succeeded in synthesizing several kinds of amino acids. Since then experiments have been performed using more oxidized gases, such as a mixture of CO, CO_2, and N_2, but it is extremely difficult to synthesize amino acids from such matter. Thus from a biochemical viewpoint a reducing primitive atmosphere is expected when explaining the origin of life. Prior to Miller's experiment, H.C. Urey had claimed that the main components of the primitive atmosphere were probably CH_4, NH_3, H_2O, and H_2, based on geochemical considerations. The gas mixture used by Miller was also based on the atmospheric composition of the primitive earth as concluded by Urey. Miller estimated that the primitive atmosphere postulated by Urey contained more than a thousand times the present amount of H_2, that is, a gas whose elemental composition resembled the solar abundances was in a state of thermodynamic equilibrium under normal temperature. As the example of rare gases reveals, however, there was a marked deficiency of volatile components in the primitive earth compared to their solar abundances. Thus the amount of H_2 in the primitive atmosphere should have been far less than the amount assumed by Urey. Hence if a gas containing only a small amount of H_2 was in a state of chemical equilibrium under normal temperature, it would be far more oxidizing than the primitive atmosphere concluded by Urey.

As discussed earlier, the earth's atmosphere was formed through degassing from inside the solid earth. If we accept this, we should regard the composition of the primitive atmosphere as having been virtually ordained under the high temperatures inside the earth, where chemical reactions proceed rapidly. Hence, we may conclude that the specific composition of the primitive atmosphere was that of an oxidized gas resembling present volcanic gas, and that the chemical composition of its volatile component system (H, C, N, O, S, etc.), which was in a state of chemical equilibrium with rocks under a high temperature close to melting point, consisted mainly of CO_2, N_2 and H_2O. This means that the primitive atmosphere would have contained N_2 and H_2O right from the start, and that its composition resembled that of the present atmosphere, except for O_2, which has been added subsequently through biological activity.

The H_2O that degassed as vapor from inside the high-temperature earth cooled on the surface of the earth and condensed to form the first oceans. Once the oceans were formed, the CO_2 in the atmosphere would have been absorbed by the seawater, so that the composition of the atmosphere would have changed to one consisting mainly of N_2. The question of from when did a significant amount of O_2 exist in the earth's atmosphere comes down to the issue of the origin of life on earth. Considerable research has been carried out in an attempt to find traces of life in the Isua metamorphic rocks from Greenland, which are the oldest known rocks on earth (metamorphic rocks whose source rock is sedimentary rock that formed about 3800 million years ago). Reasoning from the isotopic ratio of C, M. Schidlowski has claimed that the C in Isua metamorphic rocks includes some C of organic origin, but these results have not been confirmed by other researchers. Isua metamorphic rocks have undergone severe metamorphism, and so their original C isotopic composition has been disturbed, making it extremely difficult to ascertain their original composition. The existence of unmetamorphosed C that is obviously of organic origin has been recognized in rocks about 3000 million years old, such as Fig Tree shale and Sudan shale. The appearance of life, however, does not immediately signify an abundant supply of O_2 in the atmosphere. In reality, large-scale banded iron formations [sedimentary layers in which layers consisting mainly of quartz microcrystals and layers of oxidized iron (Fe_2O_3)] alternate; these are seen characteristically in geological formations 1000 to 3000 million years old) are thought by many researchers to suggest that the surface of the earth at that time was considerably more reducing than nowadays. This may indicate that a sufficient quantity of O_2 had not yet been accumulated on earth at that time. A promising new approach to setting limits on the O_2 content of the Precambrian atmosphere is the study of Precambrian soil (paleosols). H.D. Holland and E.A. Zbinden concluded considerably low oxygen pressure of about 3×10^{-5} atm at mid-Precambrian time from their study on the behavior of iron in Precambrian paleosols.

3.9 The Primordial Mantle and the Differentiated Mantle

As stated in Section 3.3, analyses of seismic waves have revealed that the density distribution within the mantle changes at the depth of 400–700 km. No final conclusion has been reached as to the cause of this discontinuity, but the view that it is the result of the phase transition of olivine, a major constituent mineral of the mantle, is convincing. As far as the seismic data and the results of high-pressure experiments are concerned, it appears that the mantle has a virtually uniform chemical composition as a whole. After differentiation of the earth's crust the mantle would have a chemical composition different to that of the undifferentiated mantle – we will refer to this as the primordial mantle. Thus if the mantle does actually have a more or less uniform chemical composition as a whole, then we are forced to conclude that the whole mantle was involved in the differentiation of the earth's crust from the mantle. However, this is unlikely, as discussed below.

In general it is thought that hot spot volcanic rocks are derived directly from the deepest part of the mantle, at least deeper than the plate that constitutes the upper mantle. This has been concluded from the fact that the magma sources of hot spot volcanoes are cut off from the moving upper plate. This is one outstanding result put forward by J.T. Wilson and J. Morgan and their colleagues on the basis of the plate tectonics theory. J.G. Schilling and his colleagues, who are carrying out research into hot spot volcanic rocks, have discovered that on the whole these rocks contain more elements that are easily concentrated in the earth's crust, such as K and Rb, than does MORB. They interpreted this as meaning that an undifferentiated primitive mantle remains in the deeper part of the mantle, and that hot spot volcanoes are derived from this deep, undifferentiated mantle, so that they contain in abundance elements that are relatively highly volatile. At present it is agreed that volcanoes in Hawaii, Iceland, the Reunion Islands, and many other oceanic islands are of hot spot origin, as well as Yellowstone in continental America.

The isotopic composition of the He contained in volcanic rocks suggests the same conclusion. With the exception of diamonds, the He isotopic ratios ($^3He/^4He$) currently observed in terrestrial material are all smaller than 5×10^{-5}, more than an order of magnitude smaller than the primordial He (solar He or planetary He: see Sect. 3.7). This is due to the increase in ^4He as the result of the decay of U and Th in the earth's interior. In the earth's crust, where U and Th are more concentrated, the value of $^3He/^4He$ is normally less than 10^{-7}. MORB, however, characteristically shows a value of $^3He/^4He \sim 1.1 \times 10^{-5}$. The reason for the figure significantly higher than that of the He in the earth's crust is simply a reflection of the smaller amount of U and Th in the mantle than in the earth's crust. The considerable uniformity of the isotopic ratios can be attributed either to the fact that the mantle region from which the MORB differentiated had quite a uniform chemical composition, or the fact that the great ease with which He moves means that it is well-mixed within the upper mantle. However, the He contained in volcanic rocks in Hawaii and Iceland often shows a markedly higher value than MORB – $^3He/^4He \cong 2 \sim 3 \times 10^{-5}$. This is generally attributed to the fact that the magma source of hot spot volcanic rocks is located at a deeper site than the magma source of MORB, and the primordial He is still preserved at this site.

Comparisons of hot spot volcanic rocks and MORB show that in terms of its chemical composition the mantle consists of at least two different regions – an undifferentiated region and a differentiated mantle remaining after the earth's crust was squeezed out. Judging from the origin of hot spots, the undifferentiated mantle lies in the deeper part of the mantle.

The existence of a primordial mantle can be demonstrated even more conclusively from observations of Xe isotopic ratios. Let us now move on to a discussion of these ratios.

Xe has nine isotopes with the mass numbers of 124, 126, 128, 129, 130, 131, 132, 134, and 136. ^{129}Xe contains isotopes produced through the radioactive decay of ^{129}I ($T_{1/2} = 1.59 \times 10^7$ years). ^{131}Xe, ^{132}Xe, ^{134}Xe and ^{136}Xe contain isotopes produced from ^{244}Pu ($T_{1/2} = 8.3 \times 10^7$ years) through fission. (Xenons produced through the spontaneous fission of ^{238}U also exist, but their contribution in the earth's interior is negligible by comparison to Xe of ^{244}Pu origin). As already stated in Chapter 2, at the time of its birth the earth trapped some ^{129}I and ^{244}Pu. Therefore, since the formation of the earth ^{129}I and ^{244}Pu have produced ^{129}Xe* and $^{131-136}$Xe^f (f stands for fissiogenic). From a comparison with the Xe in meteorites we can conclude how much ^{129}Xe* and $^{131-136}$Xe^f are contained in the Xe in the present atmosphere. Chondrites commonly contain a phase with an abnormal concentration of rare gases. This concentrated phase represents less than 0.1% of the total weight, but normally it accounts for most of the rare gases in meteorites. Except for the observation that rare gas concentrated phase is not affected by HCl and HF, we do not yet know what kind of chemical composition or crystal structure – supposing that it consists of crystals – it possesses. Nor do we have a clear understanding of how this phase came to contain such a large amount of rare gases. However, it is very likely that these rare gases were captured directly from the surrounding solar nebula when meteorites were formed, and that they are the rare gases that actually existed in the solar nebula when the solar system was formed. The rare gas components observed in this phase of concentrated gases are called "planetary gases". Comparing the Xe in the earth's atmosphere with planetary Xe, that in the earth's atmosphere contains ^{129}Xe and $^{131-136}$Xe in relatively greater abundance. This excess ^{129}Xe and $^{131-136}$Xe is the result of the incorporation of ^{129}I and ^{244}Pu at the time of the earth's birth, and this then decayed or underwent fission. From a comparison of this atmospheric Xe and planetary Xe we can conclude that about 7.4% of the ^{129}Xe and about 4.7% of the ^{136}Xe (in the discussion below we will focus only on

^{136}Xe^f as the representative of $^{131-136}$Xe^f) in the atmosphere were formed in the earth's interior from ^{129}I and ^{244}Pu respectively.

Another interesting fact about the Xe in the earth is that the Xe contained in material derived from the earth's interior, such as volcanic rocks and well gas, contains even more excess ^{129}Xe* and $^{131-136}$Xe^f than atmospheric Xe. One of the samples that has been examined in detail is the CO_2 well gas that has been collected in New Mexico in America. The Xe contained in this CO_2 well gas has been measured independently by four different groups so far, and extremely accurate Xe isotopic ratios have been obtained. The He isotopic ratios also indicate that some of the rare gases contained in the CO_2 well gas were derived from the mantle. Of particular interest here is the (^{129}Xe*/^{136}Xe^f) ratio. The (^{129}Xe*/^{136}Xe^f) ratio of the CO_2 well gas in New Mexico is approximately 4.7, significantly smaller than the value of 6.8 for atmospheric Xe. Before we discuss the geochemical significance of the (^{129}Xe*/^{136}Xe^f) ratio, let us consider again atmospheric degassing from the mantle.

Let us now consider the mantle region involved in the atmospheric degassing. We will refer to this as mantle region A. Let us compare the Xe remaining in mantle region A with the atmospheric Xe. If degassing from the mantle occurred after a sufficiently long time (t $\gg$ the half-lives of ^{129}I and ^{244}Pu) after the formation of the earth, at the time of the degassing both ^{129}I and ^{244}Pu must have already decayed from region A. Hence we can ignore changes in the Xe isotopic composition in region A after the degassing had occurred. If we suppose, however, that the degassing occurred immediately after the birth of the earth, ^{129}I and ^{244}Pu would remain in region A even after the atmospheric degassing, and would eventually decay and form ^{129}Xe* and $^{131-136}$Xe^f. As a result, region A would always contain excess ^{129}Xe and $^{131-136}$Xe compared to the atmosphere. Moreover, since the half-life of ^{244}Pu is more than five times that of ^{129}I, the later the degassing, the greater would be the contribution of ^{244}Pu-induced $^{131-136}$Xe^f relative to ^{129}Xe* decayed from ^{129}I. Summing up the above, the following relation should hold between atmospheric Xe and the Xe in mantle region A that degassed the atmospheric Xe:

$$\left(\frac{^{129}Xe*}{^{136}Xe^f}\right)_A \lesseqgtr \left(\frac{^{129}Xe*}{^{136}Xe^f}\right)_{atmosphere}. \tag{3.5}$$

However, the Xe in the CO_2 well gas in New Mexico has a significantly larger (^{129}Xe*/^{136}Xe^f) value than atmospheric Xe, and does not fulfil Eq. (3.5). This demonstrates that the region that supplied the Xe to the CO_2 well gas had no connection with the region from which the atmosphere degassed, i.e., that it was not involved at all in the formation of the atmosphere within the mantle (naturally including the formation of the earth's crust), and hence a primordial region still remains.

Anomalous Xe isotopic ratios have been detected in many other materials of mantle origin (mantle xenoliths, volcanic rocks, diamonds, etc.) in addition to the New Mexico CO_2 well gas. However, the Xe content in these samples is extremely low, and no analyses of their isotopic ratios have been carried out with sufficient precision to enable us to confirm the relation in Eq. (3.5). Consequently, it is not yet known how general in mantle-derived material are the Xe isotopic ratio anomalies that show the characteristics of the primordial mantle, or how much of the mantle the primordial mantle occupies. More precise Xe isotopic ratio measurements of mantle material in the future may produce a more concrete conclusion as to the extent of the primordial mantle.

4 Changes in the Earth's Crust

4.1 Rock Magnetism and Paleomagnetism

In the previous chapter we looked at how the isotopic compositions of various elements, particularly radiogenic elements, are effective in throwing light on the history of the earth, acting as "fossils" that record the evolutionary process of the earth. Isotopic ratios were used to trace the movement of material and elements in the earth's interior. The paleomagnetic methods to be discussed in this chapter are effective in tracing past tectonic movements of the earth's crust, such as continental drift. Here the remanent magnetization in rocks provides their location in the paleomagnetic coordinate system, and with this as a guide it is possible to track the trajectory of the movement. The remanent magnetization in rocks is used as a tectonic tracer. Paleomagnetic methods have been applied to a great number of problems in earth science since the 1950's, and have recorded striking achievements. These methods have made a particularly major contribution to establishing the foundation of plate tectonics theory, which has revolutionized earth science in the latter half of the 20th century by providing conclusive proof of the ocean floor spreading theory and the continental drift theory. In this chapter we will first explain briefly the mechanism behind rock magnetism, which is the basis of paleomagnetism, and then discuss general paleomagnetic methods. We will begin in the following section by taking a look at the history of crustal movements, which is gradually being brought to light through such research.

a) Remanent Magnetization in Rocks

Volcanic rocks usually have quite strong magnetization of about 1 A m^{-1}. This is because the minute titanomagnetite minerals contained in volcanic rocks are magnetized. Titanomagnetite is a solid solution of magnetite (Fe_3O_4) and ulvöspinel (Fe_2TiO_4). The Curie point of magnetite is $575\,°C$, but as the ulvöspinel content rises, the Curie point falls. Usually the titanomagnetite in volcanic rocks contains 40–70% of ulvöspinel, and its Curie point is about 200–400 °C.

When magnetic substances cool in the presence of a magnetic field from a temperature higher than their Curie point, they acquire very strong and stable remanent magnetization that is parallel to the direction of the external magnetic field. The resulting remanent magnetization is known as thermo-remanent magnetization (TRM). When volcanic rocks erupt as magma, the temperature of the magma exceeds 1000 °C, which is of course higher than the Curie point of titanomagnetite. So as the magma cools to form volcanic rocks, it acquires thermo-remanent magnetization that is parallel to the earth's magnetic field. As well as being strong, this thermo-remanent magnetization is also extremely stable. The remanent magnetization of volcanic rocks that cooled very rapidly after erupting and that therefore have very fine magnetic minerals (less than a few microns) is particularly stable. From experiments it has been concluded that the remanent magnetization in basalt varies hardly at all even over a period as long as the age of the earth, providing that it is maintained at room temperature. The stability of the thermo-remanent magnetization in volcanic rocks has also been supported theoretically. L. Néel, who was awarded the Nobel prize for his work in ferrimagnetism, has given a brilliant theoretical explanation of the striking stability of the thermo-remanent magnetization in volcanic rocks. A paper published by Néel in 1949, entitled *Theorie du trainage magnetique des ferromagnetiques au grains fins avec application aux ferres cuites*, retains today its value as a classic on rock magnetism. Even earth scientists not studying rock magnetism would find it worthwhile to peruse this paper, as it elucidates how to employ a physical approach in resolving earth science problems.

As long as room temperature is maintained, both theory and experiments guarantee that the thermo-remanent magnetization of volcanic rocks will be preserved stably even over a period as long as the age of the earth. This raises the possibility that the remanent magnetization in volcanic rocks may provide a faithful record of the geomagnetic field tens or hundreds of millions of years ago. This suggests that the remanent magnetization of volcanic rocks can be regarded as a faithful fossil of the geomagnetic field in the past. In addition to its stability and strength, thermo-remanent magnetization also has the major merit that it is possible to identify clearly the time at which it was acquired. In general, the time at which the remanent magnetization of volcanic rocks was acquired corresponds to the time at which the magma cooled and solidified. Consequently, the age of volcanic rocks is also the time at which the thermo-remanent magnetization was acquired, enabling us to estimate the age of the "geomagnetic field fossils".

Thermo-remanent magnetization has many outstanding features as a fossil of the past geomagnetic field, but one of its weak points lies in the intermittent nature of the eruptions of volcanic rocks. Even on a global scale, volcanic eruptions occur only intermittently, and so it is impossible to trace continuous changes in the geomagnetic field in the past if only volcanic rocks are used as samples. Let us next consider the remanent magnetization in sediment as a sample providing us with a continuous record.

Sediment is formed throughout geological time by being continuously deposited on the ocean floor. On the deep sea floor sedimentation proceeds at the rate of a few millimeters or less every 1000 years. Hundred-meter long cores [sediment is normally collected as a cylinder (core) several centimeters in diameter] collected from the sea floor will easily cover a period of 100 million years. The magnetic minerals contained in this deep sea sediment carry remanent magnetization. In contrast to volcanic rocks, the composition of these minerals resembles that of pure magnetite, and so their Curie point is also close to 600 °C. Various theories have been put forward as to the origin of this magnetite, such as that it was drawn from rocks on land – mainly granite – or that it was formed chemically through a reaction with seawater in submarine volcanoes and in sediment, or more recently an argument claiming that it is of biogenic origin, but no final conclusion has been reached. In the past the mechanism by which remanent magnetization was acquired in sediment was explained as a process by which magnetic particles slowly sinking in the seawater align themselves along the direction of the geomagnetic field, and then deposit to form sediment. Today, however, almost no scientists accept this simplistic mechanical theory. Most researchers believe that remanent magnetization was acquired slowly over a period of thousands of years. Many are of the opinion that compression after sedimentation probably played an important role in acquiring remanent magnetization. As can be inferred from this mechanism, uncertainty of several thousands or tens of thousands of years remains in assigning the acquisition time of the remanent magnetization In nearly all cases, however, the dating of sediment is based on microfossils, and compared to the uncertainty of this fossil age, the uncertainty in the acquisition time of remanent magnetization can be ignored. The remanent magnetization of sediment is known as depositional remanent magnetization or DRM.

Measurements of the remanent magnetization in deep sea sediment boring cores of several hundred meters can be expected to provide a continuous record of the geomagnetic field over more than 100 million years. In actual practice, however, several difficulties must be overcome before we can deduce the past geomagnetic field from measurements of the remanent magnetization of sedimentary rocks. In the first place, remanent magnetization is very weak, so special magnetometers are necessary. A cryogenic magnetometer using a Josephson element, which came into widespread use in the 1970's, solved this problem. But the weakness of the remanent magnetization in sedimentary rocks leads to another serious problem. Normally, when magnetic substances are left in a magnetic field they are gradually magnetized in the direction of the external magnetic field. This is known as viscous remanent magnetization or VRM. Rocks that have existed under the earth's magnetic field have all come under its influence and acquired this viscous remanent magnetization. In volcanic rocks and other rocks that have very strong thermo-remanent magnetization, however, the viscous magnetization acquired later is far weaker than the thermo-remanent magnetization, so it can be virtually ignored. Consequently, there is little obstacle to regarding the remanent magnetization of volcanic rocks as corresponding to the thermo-remanent magnetization acquired when the rocks formed. However, there are also cases in which weak depositional remanent magnetization is of about the same strength as the viscous remanent magnetization, and so the remanent magnetization of sedimentary rocks can no longer be regarded as a fossil of the past geomagnetic field. Rather, in many cases the magnetization in sediment should be regarded as a composite of the primary magnetization acquired at the time of sedimentation and the viscous remanent magnetization that has been gradually acquired under the geomagnetic field ever since the time of sedimentation. In order to isolate the primary remanent magnetization that is useful as a record of the past geomagnetic field, we must remove the unnecessary secondary remanent magnetization. The alternating current demagnetizing method is used for this. Sedimentary samples that have undergone careful alternating current demagnetizing provide an unparalleled record of variations in the geomagnetic field over a long period.

b) The History of the Geomagnetic Field

Figure 4.1 shows the changes in the polarity of the earth's magnetic field as found from deep sea cores. As the figure reveals, the geomagnetic field has constantly reversed its polarity throughout geological time. We still have no proper understanding of the mechanism by which these reversals of the geomagnetic field occur. Normally remanent magnetization tends to weaken at around the time of these reversals. Some scientists speculate, therefore, that immediately before the geomagnetic field reverses it virtually disappears temporarily, and when it recovers it grows in the opposite direction purely by chance. It is not difficult to imagine that a complete disappearance of the geomagnetic field, even if only temporary, would have a considerable influence on the earth's environment. For instance, the solar wind and the charged particles in cosmic rays are deflected by the geomagnetic field and prevented from penetrat-

Rock Magnetism and Paleomagnetism

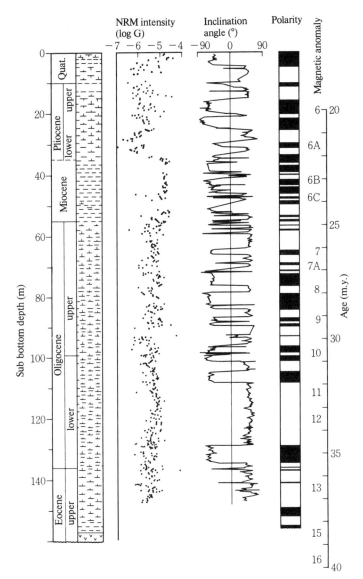

Fig. 4.1. Paleo-geomagnetic field recorded in a deep-sea sediment core. The sample is from DSDP (Deep Sea Drilling Project) Leg 73 drilled core from southern Atlantic Ocean. Intensity of remanent magnetization is in the unit of Gauss (shown as Log G in the 3rd column). (After Tauxe et al., 1980)

ing directly to the earth, but if the geomagnetic field disappeared, these charged particles would enter the upper atmosphere directly. It is not clear to what extent this would affect the temperature and other aspects of the climate, but some researchers link reversals of the geomagnetic field to global changes in the climate.

At present the oldest deep sea sediment available for study is no more than 200 million years old at the most. This is because the old (more than 200 million years old) oceanic crust has subducted into the mantle and disappeared, and the present sea floors are all relatively young (less than 200 million years old). So we cannot seek a continuous record of the geomagnetic field back beyond this time. It is not clear whether reversals of the geomagnetic field occurred frequently in Mesozoic times more than 200 million years ago and also in the Precambrian period. As far as we can tell from the intermittent records using volcanic rocks and sedimentary rocks, however, it seems that reversals of the geomagnetic field occurred then in more or less the same manner as they have ever since Mesozoic times.

The remanent magnetization in volcanic rocks is strong and stable, and in most cases it can be regarded as a faithful record of the past geomagnetic field. With depositional remanent magnetization, however, the effect of viscous remanent magnetization cannot be ignored, and considerable care is necessary in handling this. However, sediment does have the major merit of providing a continuous record of the geomagnetic field, and is thus an indispensable sample in tracing its history. Combined with this, we can obtain more reliable data on the past geomagnetic field by using as checkpoints the remanent magnetization in volcanic rocks at several points in time. However, the thermo-remanent magnetization in volcanic rocks provides us with the sole reliable data when discussing not only the past direction of the geomagnetic field, but also its intensity.

The measured values available to us as experimental data on the remanent magnetization of volcanic rocks are the dip (written as I; denoting the inclination or downward angle from the horizontal plane as positive) and the declination (written as D; denoting the north pole as 0 and the eastern angle as positive) of the site at which the sample was collected. From these two data (I, D) on remanent magnetization we can deduce the geomagnetic field that formed this remanent magnetization, but before discussing that it is necessary to add a few explanatory notes about the general expression "geomagnetic field".

The present geomagnetic field can be likened to that created if a permanent magnet (magnetic dipole) were placed in the center of the earth with the N pole facing southwards. This is known as the geocentric dipole magnetic field. The resemblance of the geomagnetic field to the dipole magnetic field was first proved by C.F. Gauss, the great 19th century German mathematician, using the mathematical method that he had invented, known as spherical harmonic analysis. The current geomagnetic field is about 80% dipolar – the main geomagnetic field – and about 20% nondipolar. The present geomagnetic field dipole is inclined at an angle of about 11.5° from the earth's axis of rotation and is precessing around the axis of rotation in a period of tens of thousands of years. Consequently, if we average the geomagnetic field over more than a few ten thousand years, the geomagnetic dipole axis is expected to coincide with the rotation axis of the earth. This is in fact the case at least for the last few million years; the geomagnetic dipole axis averaged for the Quaternary and the late Tertiary volcanic rocks shows good coincidence with the earth's rotation axis. Let us suppose now that the geomagnetic dipole is located in the center of the earth, and that it is not parallel to the earth's axis of rotation. The points at which the dipole intersects with the surface of the earth when extended are called the geomagnetic poles. Currently (1975) the north geomagnetic pole is located at 78.6° N, 70.5° W and the south geomagnetic pole at 78.6° S, 109.5° E. Note that the geomagnetic poles differ from the magnetic poles at the point where the dip is ($\pm$) 90°, i.e., where the magnetic needle stands perpendicular to the surface of the earth. This is due to the nondipole component. Now let (I,D) be the components of the geomagnetic field measured at the observation point (latitude: λ_0, longitude: ϕ_0). Using spherical geometry, we can calculate the the location (latitude = λ, longitude = ϕ) of the geomagnetic pole corresponding to the geomagnetic dipole which produces the observed geomagnetic field at (ϕ_0, λ_0):

$$\sin \phi = \sin \phi_0 \cos p + \cos \phi_0 \cdot \sin p \cdot \cos D$$
$$\sin (\lambda - \lambda_0) = \sin p \cdot \sin D / \cos p. \qquad (4.1)$$

Here

$$\cot p = (1/2) \cdot \tan I.$$

Let us now discuss how to find the past geomagnetic field (paleomagnetic field) based on the remanent magnetization in volcanic rocks. Let us suppose for now that volcanic rocks collected at the site (λ_0, ϕ_0) record faithfully the direction of the geomagnetic field at the time when the rocks erupted to the surface of the earth t years ago. By inserting in Eq. (4.1) the direction of the remanent magnetization measured in the volcanic rocks, we can find the location (λ, ϕ) of the geomagnetic pole of the geomagnetic field t years ago. The poles found in this manner are called the virtual geomagnetic poles, or VGP.

As already stated above, the geomagnetic dipole is precessing around the earth's axis of rotation in a period of tens of thousands of years. In addition to this precession, the geomagnetic dipole also moves even

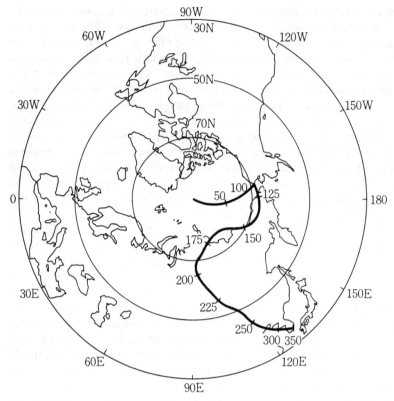

Fig. 4.2. Polar wandering curve determined from North American samples. *Numbers* on the curve indicate an age (millions of years). (After Irving, 1979)

more slowly on an even longer time scale. Figure 4.2 shows an example of this. Results based only on samples from North America are shown here. Note that as the age increases, the VGP's become further distant from the current geomagnetic pole. In Fig. 4.2 the VGPs found from samples of different ages are joined in chronological order. This curve is known as the polar wandering curve, and as will be explained later, it is the most fundamental piece of data when using paleomagnetic data to trace crustal movements, i.e., paleomagnetism in a narrow sense. Before moving on to a discussion of paleomagnetism, let us comment on a few results of geomagnetic field research that are of interest from a geohistorical viewpoint.

As in the next section, when discussing paleomagnetism it is always assumed that the geomagnetic field has been a geocentric dipole field parallel to the earth's rotation axis. I have already stated that the present geomagnetic field is close to the geocentric dipole field. Naturally this conclusion is based on observed data from geomagnetism observatories around the world. However, these data do not cover even the past century. The geomagnetic field in earlier periods must be examined by using the remanent magnetization in rocks as the lead. So far the results of spherical harmonic analyses of paleomagnetic samples up to several millions of years ago indicate that the supposition that the geocentric dipole is parallel to the earth's axis of rotation (which is called the geocentric axial dipole field) is quite valid. We have no clear proof, however, that the geomagnetic field prior to that can be regarded as corresponding to the geocentric dipole, as is the case today. When treating a period longer than several tens of millions of years, unknown factors other than remanent magnetization come into play, such as crustal movements, so that even if the remanent magnetization itself is stable, it can no longer be regarded as a fossil of the past geomagnetic field. However, by assuming that in the geomagnetic field more than tens of millions of years ago the geocentric dipole was parallel to the earth's axis of rotation, and by using paleomagnetic data as the tracer of crustal movement, valid earth science conclusions, e.g., the continental drift to be discussed in a later section, can be reached in many cases. Hence it seems reasonable to regard the assumption of a geocentric axial dipole as perfectly appropriate at least up until about 200 million years ago when continental drift commenced.

So we can regard the geomagnetic field as having had a geocentric axial dipole at least over the past several hundred million years. How strong was the geomagnetic field? As well as its stability and the fact that it lies parallel to the external magnetic field, thermo-remanent magnetization is characterized by the fact that its intensity is proportional to the external magnetic field. This property can be used to estimate the strength of the geomagnetic field in the past. First the strength of remanent magnetization in volcanic rocks is measured, then these rocks are heated above their Curie point and cooled under the present magnetic field. The intensity of the resulting artificial thermo-remanent magnetization is then measured. The ratio between the two is equal to the ratio between the past and present intensity of the geomagnetic field. This experiment, however, runs into many practical difficulties, including the fact that samples undergo alteration when heated in the process of creating artificial thermo-remanent magnetization. E. & O. Théllier devised a highly ingenious method, known as the Théllier method, for overcoming this difficulty. It would be no exaggeration to say that nowadays the only meaningful estimates of the paleomagnetic field intensity are ones using the Théllier method. Note, however, that even though the Théllier method is a necessary condition, it is not a sufficient condition. Thus estimates of the intensity of the paleomagnetic field involve extremely difficult and time-consuming experiments, and pro-

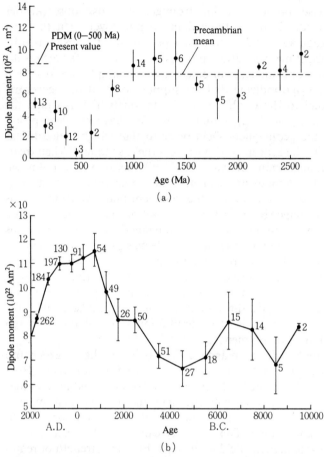

Fig. 4.3. Temporal variation of geomagnetic dipole moment. **a** From the present to about 2700 million years ago. *Dotted lines* indicate average values for Precambrian period (650–3000 million years) and for the last five million years. PDM for paleo dipole moment. **b** From the present to about 10,000 B.C. *Numbers* indicate number of samples used for measurement. (After McElhinny and Senanayake, 1982)

duce few results. To briefly sum up the results obtained so far, there is no evidence that the strength of the geomagnetic field over at least the past several hundred million years differed significantly from the present value of approximately (5×10^{-5} T). Figure 4.3a shows the changes in the intensity of the earth's magnetic field throughout geological time. The data use the thermo-remanent magnetization in volcanic rocks.

By contrast, fragments of baked earths from about 10 000 years ago, when traces of mankind appear, provide excellent paleomagnetic data. The main magnetic material in baked earth is hematite (γ Fe_2O_3), which is even more oxidized than magnetite, so that even if artificial thermoremanent magnetization experiments are performed, it is far less likely that the samples will undergo thermal alteration than in the case of volcanic rock. Using baked earth as the sample, Fig. 4.3b shows the curve for variations in the intensity of the geomagnetic field. Note that the geomagnetic field approximately 6000 years ago was about half as strong as it is today.

c) The Origin of the Geomagnetic Field

When was the geomagnetic field first produced on the earth? According to current geomagnetic field theory, the earth's magnetic field is formed by fluid movement within the core of the earth. This idea goes back to the dynamo theory of a theoretical physicist, J. Larmor, in 1919. Larmor proposed the dynamo theory to explain the origin of the solar magnetic field. According to this theory, a fluid with good electrical conductivity circulates in the sun's interior, and this movement interacts with the magnetic field that first existed to eventually create a constant magnetic field. The name dynamo theory was derived from the fact that this mechanism is the same as that of actual electric dynamos. The fluid movement substituted instead of the dynamo rotor corresponds to the dynamo producing the magnetic field. A constant current entails a constant magnetic field. E.C. Bullard and W.M. Elsasser later applied the dynamo theory to explanations of the origin of the geomagnetic field, and this theory has subsequently undergone theoretical refinements in the hands of many researchers, so that nowadays it is virtually the only theory on the origin of the earth's magnetic field. Although the outline of the dynamo theory is generally accepted, some details remain unsettled. Of particular importance is the issue of the energy source behind this fluid movement within the core. At one time it was commonly attributed to the energy produced through the radioactive decay of such elements as U, Th, and K, which are thought to be contained to a certain extent in the core. However, many researchers question whether a sufficient amount of radioactive elements to cause thermal convection actually exists in the metallic core, which consists mainly of Fe and Ni. In recent years the theory that the fluid movement within the core is caused by the energy released when the inner and outer cores separated has gradually been gaining ground. According to this theory, when the core was first formed, consisting mainly of Fe and Ni as well as about 20% of light elements (at present it is not clear what these elements are), it was in a molten state, and Fe and Ni later condensed and settled down to the center of the core, where they solidified and gradually formed an inner nucleus.

Setting aside the details of the dynamo theory, the geomagnetic field is intrinsically related to the existence of the earth's core. In this book we have considered the earth's core as having formed at more or less the same time as the earth formed. Based on this viewpoint, there is no objection to seeking the origin of the geomagnetic field in the very early stages of the earth's evolution.

Dacite rocks from Western Australia that have been studied by W.M. McElhinny and W.E. Senanayake are thought to have formed in the early Precambrian period and to almost certainly have primary remanent magnetization. The U–Pb age found for the zircon contained in these rocks is 3452 ± 16 million years, placing them among the oldest rocks on the Australian continent. The remanent magnetization contains components that are very stable against the alternate current demagnetization and thermal demagnetization, and this is probably the primary remanent magnetization that was acquired when the rock was formed approximately 3500 million years ago. Thus it seems that the origin of the geomagnetic field goes back at least 3500 million years.

The oldest known rocks on earth, the Isua metamorphic rocks from Greenland, have quite stable and strong remanent magnetization. It is difficult, however, to determine whether or not this was acquired at the time of metamorphism. Judging from its stability, however, it seems closer to thermal remanent magnetization than to secondary viscous magnetization. If the remanent magnetization in the Isua metamorphic rocks is of primary origin, it will mean that the existence of the geomagnetic field dates back to 3800 million years ago. Since no rocks older than this can be found on earth, however, we cannot reach a clear conclusion about the state of the geomagnetic field prior to that time.

4.2 Ocean Floor Spreading, Continental Drift, and Plate Tectonics

Striking progress has been made in earth science since World War II, and the results of research into the oceanic crust are particularly remarkable. These results developed into the ocean floor spreading theory and then the plate tectonics theory. Since the 1960's, plate tectonics has been a central theme in solid earth science, and countless papers have been published on this subject. This field covers a broad spectrum of issues, and it is impossible here to discuss all of these, nor is that the purpose of this book. Here we will limit our discussion to the most fundamental aspects of plate tectonics, such as continental drift and ocean floor spreading, looking at these from the viewpoint of geohistory, which traces the overall development of the earth.

Shortly after World War II many studies of the ocean floor structure of the Atlantic and Pacific Oceans were carried out using acoustic echo sounding and magnetic surveys, with the United States and Britain taking the lead. The first major result of these studies was the discovery of enormous mountain ranges on the floor of these oceans, far surpassing any mountains on the continents. Most typical is the Mid-Atlantic Ridge running more or less along the center of the Atlantic Ocean from north of Iceland down to the Antarctic Ocean in the south. These enormous mountain ranges on the ocean floor, known as mid-ocean ridges, are found not only in the Atlantic Ocean, but also in the Pacific and Indian Oceans, and are now regarded as among the most fundamental structures on the ocean floor.

At about the same time as the discovery of mid-ocean ridges, the results of magnetic measurements by observation vessels brought to light another interesting discovery – the existence of magnetic anomaly lineation. Here "magnetic anomaly" can be simply regarded as a deviation from the average geomagnetic field (strictly speaking, the magnetic field determined as the international standard on the basis of data from observation stations around the world). Most ocean floors have what is known as magnetic anomaly lineation, where bands of positive and negative magnetic anomaly regions alternate. As the name indicates, the magnetic anomalies form an impressive banded pattern. Of further interest is the fact that these banded patterns spread out more or less symmetrically on either side of the mid-ocean ridges. F.J. Vine and D.H. Matthews have put forward the following explanation for these magnetic anomaly bands. Magma is constantly gushing out on to the ocean floor at the mid-ocean ridges, where it cools and forms the oceanic crust. Magma welling up subsequently then pushes the oceanic crust out on either side at right angles to the mid-ocean ridge. This is the basic outline of the ocean floor spreading theory. Magma that has cooled on the earth's surface acquires thermo-remanent magnetization parallel to the direction of the geomagnetic field at that time. This means that the remanent magnetization switches direction in line with changes in the polarity of the geomagnetic field. So with the mid-ocean ridge as the axis of symmetry, the oceanic crust formed there is magnetized in alternating directions to form a banded magnetic pattern that is known as the magnetic anomaly lineation pattern. A similar explanation was proposed independently by L. Morley and A. Larochelle at about the same time as the Vine-Matthew theory was put forward.

Not only does this hypothesis give a very neat explanation of the magnetic anomaly bands observed on ocean floors around the world, but it also provides powerful support for the independently proposed ocean floor spreading theory, which suggests that ocean floors are formed at the mid-ocean ridges and spread out from there. The ocean

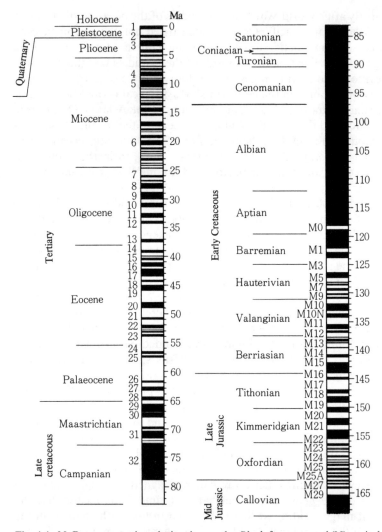

Fig. 4.4. N–R geomagnetic polarity time scale. *Black* for a normal (N) period (same as the present polarity) and *white* for a reversed (R) polarity. (After Cox, 1982)

floor formed by spreading out on both sides of the mid-ocean ridge can be expected to become older gradually as the distance from the mid-ocean ridge increases. Oceanic crust ages estimated mainly from fossils in deep sea sediment fully support this prediction. Moreover, by comparing paleomagnetic data with the geomagnetic field N–R polarity time scale (Fig. 4.4) found from the magnetic anomaly pattern, the age of individual magnetic anomaly bands can be found. In effect, the results of magnetic measurements at sea enable us to estimate the time at which the ocean floor was formed, based on the magnetic anomaly band pattern-age relation.

The observed fact that the ocean floor becomes older systematically in proportion to the distance from the mid-ocean ridge indicates the correctness of the basic framework of the ocean floor spreading theory so strongly that there is virtually no room for disagreement. Numerous attempts have been made to find more direct proof of the Vine-Matthews hypothesis. For example, a borehole several hundred meters deep was drilled directly into the oceanic crust off Bermuda, and the remanent magnetization of the rocks collected there was measured. According to the Vine-Matthews hypothesis, oceanic crust at one location can be expected to have the same age and to be magnetized in the same direction as the contemporaneous geomagnetic field. Actual measurements, however, betrayed these expectations. The lower half of the boring core was magnetized in a positive direction, and the upper half in the opposite direction. In subsequent deep sea boring conducted at different locations, many cases have been found in which the direction of magnetization of the oceanic crust changes its polarity several times as the depth increases. The current explanation of why the crust is magnetized in this manner is that some time after the oceanic crust was formed at the mid-ocean ridge – at that time the crust would have been uniformly magnetized – the polarity of the geomagnetic field must have changed, and volcanic rocks intruded the crust at that time and formed a sill magnetized in the opposite direction to the surrounding crust. The question then arises of why sill intrusion occurred successively over a period long enough to record several reversals of the geomagnetic field, and why the initial magnetic banded pattern has been preserved intact despite this large-scale sill intrusion. It seems difficult to explain these phenomena in line with the straightforward Vine-Matthews hypothesis.

Let us give another example. According to the ocean floor spreading theory, the oceanic crust ages as it moves away from the mid-ocean ridge. The oldest oceanic crust known at present is the area located immediately to the east of the Ogasawara-Mariana trench in the western Pacific, the most distant part from the East Pacific Rise (EPR). R. Larson analyzed the magnetic anomaly patterns of this region in detail, and concluded that they belong to the Jurassic period. In an effort to find evidence for the Vine-Matthews hypothesis, the International Phase of Ocean Drilling (IPOD) has carried out drilling of the oceanic crust in the magnetic anomaly zone belonging to the Jurassic period, and has succeeded in excavating bed rock from a depth of approximately 1000 m. However, the rock samples were magnetized in completely the opposite direction to that expected from the magnetic anomalies observed at sea. Also curious was the fact that K–Ar dating of these rock samples (basalt) showed that they were less than 120 million years old, far younger than the Jurassic (140 million-approx. 200 million years ago) age expected from the magnetic anomaly pattern. These K–Ar ages were found using the ^{40}Ar–^{39}Ar isochron method (see Section 4.4). Judging from the neat isochron formed, it is difficult to attribute the differences in these ages to an error in dating. The fossils in the sediment on top of the bed rock are all younger than the Cretaceous period (i.e., less than 140 million years old). Since the rock was excavated from a depth of only about 1000 meters, it was suggested that the oceanic crust belonging to the Jurassic period exists further down below this, and that the upper layer was probably formed by magma from local submarine volcanoes erupting over a wide area at quite a distance from the mid-ocean ridge. From analyses of several magnetic anomaly patterns, however, it has been estimated that the oceanic crust (magnetized) that is the cause of the magnetic anomalies at sea is less than several kilometers thick. In that case, it is difficult to suppose that the magnetized sill in the upper 1000 m or so of the magnetized layer only a few kilometers deep was formed secondarily without any relation at all to the magnetic anomaly pattern. Moreover, at this stage no evidence based on dating has been obtained to prove whether or not the strata below a depth of 1000 m really belong to the Jurassic period.

Here we have cited two examples of attempts to find direct proof for the Vine-Matthews hypothesis. Both examples show that a facile application and interpretation of this hypothesis do not accord with reality. The ocean floor spreading theory and the subsequent plate tectonics theory have successfully provided a unified explanation for numerous geological phenomena that we had previously been forced to explain in a fragmentary and somewhat arbitrary manner. Probably this is the first time in the whole history of geology that so many apparently unrelated geological phenomena, especially tectonic phenomena, have been explained by one hypothesis. It would hardly be exaggerating to say that plate tectonics provided the opportunity for a qualitative turning point in earth science corresponding to the appearance of quantum mechanics in physics. Setting aside the general outlines of the ocean floor spreading theory and plate tectonics, however, we should keep in mind the fact that many unsolved problems remain when it comes down to applying these theories to individual issues and interpreting them on this basis. The two cases of deep sea drilling described above were both carried out more than a decade ago. At that time many researchers took quite a serious view of the results as contradicting the fundamental requirements of the ocean floor spreading theory and plate tectonics. However, the subsequent spectacular success of plate tectonics – in other fields unrelated to the interpretation of these problems – meant that the seriousness of the problem gradually faded, and it now seems to have disappeared from memory. We should be fully mindful, however, that so far no definite solutions whatever to these problems have been proposed.

Although the plate tectonics theory and its forerunner, the ocean floor spreading theory, still contain many unsolved issues, there is virtually no room for doubt about their general outline, and they can be regarded as having been proved. The strongest corroboration has been provided by the oceanic crust ages that are proportional to the distance of the oceanic crust from the mid-ocean ridge. The relative movement between the American and African continents, which has been concluded from paleomagnetic results, also validates this hypothesis. The paleomagnetic approach which led to the conclusion of continental drift is built on the basis of clear-cut physics, and there is little ambiguity in how its logic is put together, so it is the most convincing of all geological arguments. We will discuss this below.

In the previous section we discussed the paleomagnetic polar wandering curves which are the most fundamental premise in the paleomagnetic approach. These were defined as the trajectories drawn on the surface of the earth by the VGP's. The present VGP's more or less agree with the astronomical poles (i.e., the earth's axis of rotation), but as we go further back into the past they gradually come to deviate from the present astronomical pole. We considered the geomagnetic field as a geocentric dipole field, and defined the VGP's as the trajectories drawn on the earth's surface. When a specific point in time is considered, the VGP's found [using Eq. (4.1)] at any position on the earth's surface are located at specific sites corresponding to that point in time. So the paleomagnetic polar wandering curve drawn by connecting in chronological order the VGP's found in this manner will be defined as a single curve on the earth's surface. In the 1950's a group led by S.K. Runcorn collected rock samples in South America and on the African continent, and found the paleomagnetic polar wandering curves based on the remanent magnetization in these samples. Figure 4.5 shows the results compiled by K.M. Creer. To their surprise, there was an obvious discrepancy between the paleomagnetic polar wandering curve found from samples gathered in South America and that found from African samples. This discrepancy is particularly striking for the VGP's before the Cretaceous period. Runcorn and his colleagues concluded that this discrepancy can be attributed to relative drift between the South American and African continents. They suggested that what should rightfully be a single paleomagnetic polar wandering curve has ended up as separate curves owing to the relative drift of the two continents.

Next they arbitrarily moved the two continents around in an attempt to determine whether or not the paleomagnetic polar wandering curves found for the two continents would overlap to form a single curve. As

Fig. 4.5c shows, when the two continents are moved in relation to each other the two paleomagnetic polar wandering curves for the pre-Cretaceous period overlap almost perfectly. Even more surprising is the fact that the relative positions of the continents at this time coincide perfectly with the pre-continental drift positions formerly proposed by A. Wegener – the coastlines of the two continents fit together snugly to form a single unit in which the South American continent nestles in the bosom of Africa. As shown in Fig. 4.5c, however, ever since the Cretaceous period the paleomagnetic polar wandering curves of the two continents have instead moved away from each other. This indicates that up until the Cretaceous period the two continents were joined together as one, but then began to separate, i.e., continental drift commenced.

The relative movement of the South American and African continents and the time of this drift, as concluded from paleomagnetic data, fit in perfectly with the ocean floor spreading theory concluded from the magnetic anomaly patterns on the sea floor and from the ages of the oceanic crust.

As a more recent example of the enormous success of the paleomagnetic approach, we may mention the development in accretion tectonics, which shows that exotic terrains are moved large distances and then accreted onto continental margins. Some researchers even argue that the growth of large continental blocks occurs mainly through such accretion of exotic terrains (For further information of the accretion tectonics, readers may refer to an excellent review by Jones et al. 1983).

As the above example shows, paleomagnetism offers a splendid means for probing past crustal movements. Here we have drawn the polar wandering curves from the remanent magnetization vectors of rocks, and concluded the relative drift of continents by comparing these curves. An enormous volume of paleomagnetic data covering a long geological span is necessary for this, but there is little room in this conclusion for ambiguity.

It should be noted here that when applying paleomagnetic methods to studies of tectonic movements, usually the discussion is based only on a comparison of the directions of the remanent magnetization (in many cases merely from its horizontal components). The remanent magnetization in rocks can be likened to arrows carved into the rocks. If the directions of the remanent magnetization in two adjacent tectonic structures formed in the same period are not parallel, we can conclude that these two structures have moved relative to each other. As long as the same point in time is being considered, the direction of the geomagnetic field at adjacent locations should be more or less parallel. However, it would be dangerous to follow the above example explaining continental drift and conclude relative movement of the two tectonic structures by overlapping the directions of their remanent magnetization. This is be-

Ocean Floor Spreading, Continental Drift, and Plate Tectonics 117

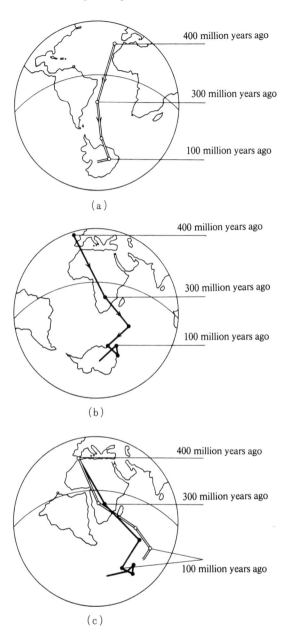

Fig. 4.5. Polar wandering curves (also see Fig. 4.2.) obtained from **a** African samples (*open circles*) and from **b** North American samples (*solid circles*). Note that both curves coincide down to about 100 million years when North America is fitted to Africa **c**. (After Creer, 1967)

cause if it is possible that the tectonic structures have moved not only within a horizontal plane, but also in a three-dimensional manner, such as tilting movements, then the relative movement found merely by overlapping the directions of the remanent magnetization will no longer be unique. In the paleomagnetic method seen in the example of continental drift, the VGP's were found by dealing not only with the horizontal components of remanent magnetization, but also the three-dimensional vectors. It should also be noted that crustal movement was concluded under constraints in which the temporal coordinates of the paleomagnetic polar wandering curves have been added, and so the conclusion as far as the relative movement of the earth's crust is concerned is virtually irrefutable.

If there is scarce room for doubt as to continental drift, what then is the moving force behind this? It is customary to assume that convection within the mantle is the cause of continental drift. However, owing to difficulties in estimating the physical properties of the earth's interior, especially the viscosity, none of the mantle convection models proposed so far is conclusive. One promising recent development, however, is seen in the attempt to estimate mantle viscosity from the post-glacial rebound of a continent. From the post-glacial uplift of the North American land mass due to the removal of the overlying ice sheet, W.R. Peltier obtained a value of approximately 10^{21} Pa s^{-1} for the viscosity in the whole mantle. The resulting Rayleigh number (a nondimensional parameter to describe the occurrence of convection, inversely proportional to viscosity) shows convincingly that there must be convection in the mantle. With better constraints on other physical properties together with a larger capacity computer, we may eventually depict the mantle convection. In parallel with these lines of research, effort in recent years has been oriented toward a discussion of the driving force behind the plates, as this is closely connected to the driving force behind continental drift. The main mechanisms that have been proposed as the driving force behind plate movements are (i) the force pushing from the mid-ocean ridges; or conversely (ii) the force that drags the plates down at the ocean trenches owing to their own gravitational force; or (iii) mantle convection. (ii) is regarded as the strongest of these forces. For example, S. Uyeda has concluded that plate movements are the result of the cooled and heavy plates – when mantle material cools on the earth's surface it becomes heavy – subsiding at the ocean trenches through their own gravity, and so the rate at which the plates move is equivalent to the final velocity of the free fall within the mantle material as the result of the plate's own gravity.

4.3 Exchange of Material Between the Mantle and the Earth's Crust

Mantle material that has extruded at a mid-ocean ridge spreads out on either side to form the oceanic crust, and eventually subsides into the ocean trenches along the continents and returns to the mantle. This sinking of the plate is referred to as subduction. The introduction of the ocean floor spreading theory sparked off a great number of studies of mid-ocean ridges. By contrast, however, it is only quite recently that systematic research has been carried out into ocean trenches. As already stated, there are powerful arguments in favor of the view that the driving force behind plate movements lies in the ocean trenches. Ocean trenches are also a very interesting site from a geohistorical viewpoint, as exchanges of material between the mantle and the earth's crust occur here. Let us now focus our discussion on these material exchanges.

After welling up at a mid-ocean ridge, mantle material forms the ocean floor and eventually subsides into the ocean trenches. If that were all, it would mean that material derived from the mantle merely returns there, and that there is no material interaction between the mantle and the earth's crust. However, on top of the oceanic crust that was born at the mid-ocean ridge and moved to the ocean trenches over a long span of time lies sediment in the form of land material carried to the sea through erosion, or the remains of living matter that had existed in the ocean. Sediment builds up on the ocean floor at the extremely slow rate of a few mm every thousand years. Thus near the mid-ocean ridge the ocean floor has a very thin sedimentary layer, while in the vicinity of the ocean trenches, which it has taken the oceanic crust more than 100 million years to reach, sediment is built up over a depth of several hundred meters. If when the oceanic crust subsides into the trench it takes this sediment down with it, the mantle will receive a continual supply of crustal material through the subsiding sediment, and this will constitute an important mechanism that cannot be ignored when considering exchanges of material between the earth's crust and the mantle. At one time it was predicted that since sediment is less dense than the bedrock on the ocean floor and since it is also soft mechanically, sediment would not subduct with the ocean floor, and would instead be scraped upwards onto the land side. In fact, it is known that the land side of the Mexican trench, for instance, forms a structure – referred to as accretionary prism – whereby ocean sediment is accumulated as if it had been scraped up onto the land side. As ocean trench surveys progressed, however, it became clear that almost no sediment (carried from the ocean side) exists on the land sides of the Mariana, Japan, or Central American trenches. It is thought that at these trenches sediment subsided into the mantle together with the subducting ocean floor. So far no convincing conclusion has been reached as to what causes the difference between these two kinds of ocean trench – ones where sediment is merely scraped up on the land side and ones in which it is dragged down into the mantle. From a practical viewpoint, some scientists are interested in ocean trenches that drag down sediment as a possible site for disposing of radioactive waste.

From the Nd growth curve (Fig. 3.3b) using data on volcanic rocks and from the $\varepsilon_{Nd} - \varepsilon_{Sr}$ plot (Fig. 3.4), in Chapter 3 we suggested that the return of material from the earth's crust to the mantle cannot be ignored. Subduction at the ocean trenches is the main mechanism behind this return flow. Proof that sediment does actually subside at ocean trenches has also been obtained from the isotopic compositions of volcanic rocks that have erupted near ocean trenches. Let us look at volcanic rocks from Martinique Island in the Lesser Antilles as one example. The Lesser Antilles are a young island arc (formed 0–25 million years ago) in the Caribbean Sea, and it is thought that subduction is continuing at the ocean trench there. Figure 4.6 plots on a $^{143}Nd/^{144}Nd - ^{87}Sr/^{86}Sr$ diagram the Nd and Sr isotopic data measured for these volcanic rocks. The average values of the Nd and Sr isotopic ratios found for MORB and Atlantic Ocean sediment are also indicated on the same diagram. The curve in this diagram represents the case when MORB and sediment are mixed. The data on the Martinique volcanic rocks lie more or less on this mixing curve. We can interpret this as meaning that the sediment which subducted together with the plate at the nearby ocean trench has

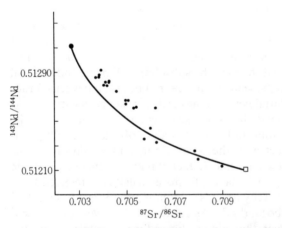

Fig. 4.6. $^{143}Nd/^{144}Nd - ^{87}Sr/^{86}Sr$ diagram. Oceanic basalt data (*small circles*) lie approximately on a mixing curve between MORB (a *large solid circle*) and ocean sediment (a *square*). (After Davidson, 1983)

mixed into the mantle below Martinique Island. However, volcanic rocks from St. Kitts, Statia and Redonda Islands, which are a little way off from Martinique Island, show Nd and Sr isotopic ratios that are almost the same as those of MORB, even though they too are from volcanoes in the Lesser Antilles. This indicates that sediment subducts quite heterogeneously even at one and the same trench. No doubt this subduction mirrors local geological structures.

The combined use of accelerators and mass spectrometers in recent years has enabled very precise measurement of the isotopic ratios of light elements such as Be and C. Using these new analytical methods, more direct evidence of the subduction of sediment has been obtained by detecting minute amounts of ^{10}Be in volcanic rocks. ^{10}Be disintegrates into ^{10}B over a half-life of 1.6 million years. ^{10}Be is being formed continuously in the upper atmosphere by cosmic ray irradiation of the upper atmosphere materials, and it then falls to earth and eventually undergoes radioactive decay and disappears. We can safely ignore the ^{10}Be production in the mantle. Thus if ^{10}Be is detected in volcanic rocks it will mean that it was originally formed in the upper atmosphere and fell to earth, where it was somehow incorporated in the volcanic rocks. L. Brown and his colleagues measured the ^{10}Be in volcanic rocks gathered from island arcs in the Aleutian Islands, Central America, the Andes, Japan, and the Mariana Islands, and succeeded in detecting significant amounts of ^{10}Be. It is difficult to attribute so much ^{10}Be to factors such as pollution by rainwater after the volcanic rocks had erupted, and so Brown and his colleagues concluded that the magma itself originally contained ^{10}Be. They explained this by saying that ^{10}Be that had formed in the upper atmosphere fell to earth and sank into the mantle at the ocean trenches along with deep sea sediment. The samples in which significant amounts of ^{10}Be were detected were all volcanic rocks from island arcs along ocean trenches, supporting this hypothesis. Almost no ^{10}Be has been detected in volcanic rocks more distant from the ocean trench area, further substantiating the theory that attributes the ^{10}Be in volcanic rocks to the subduction of sediment at ocean trenches. On the other hand, most volcanic rock samples obtained from the Andes, Japan, and the Mariana Islands, even though these belong to a subduction zone, contain only about the same amount of ^{10}Be as volcanic rocks from nonsubduction zone sources. Thus the relationship between accretional prism – the absence of this is thought to indicate the extent of sediment subduction – and other structural characteristics of subduction and the amount of ^{10}Be contained in volcanic rocks in these areas is by no means clear.

Subduction draws sediment down into the mantle. Meanwhile, the ocean floor formed at the mid-ocean ridge eventually reaches the ocean trenches after a great lapse of time. During this time the rocks on the ocean floor are in continual contact with seawater, which causes alteration. Hence basalt that has undergone alteration owing to rock-seawater interaction contains more than ten times ($H_2O = 1 - 2$ wt.%) the amount of water in basalt that has just been formed at the mid-ocean ridge. When this altered basalt subducts, a considerable amount of water is also incorporated into the mantle. E. Itoh and her colleagues have concluded that the amount of H_2O that sinks into the mantle along with the eroded oceanic crust is significantly greater than the amount of H_2O supplied to the earth's crust from the mantle through volcanic activities and in the form of hot spring water. Itoh and her colleagues have estimated the amount of this H_2O that disappears into the mantle as $1.2 - 11 \times 10^{14}$ g y^{-1}. If the upper limit of this estimate is correct, this "disappeared seawater" amounts to a considerable portion of the present amount of seawater (about 1.4×10^{24} g). This may explain the considerable lowering of the sea level (marine regression) from the Cretaceous period up to the present that has been pointed out by many geologists.

4.4 Geochronology

In the 20th century the solid earth sciences have been characterized by two revolutionary developments. One is the introduction of plate tectonics, which evolved out of the ocean floor spreading theory described in the previous section. The other is the establishment of geological dating methods using radioactive isotopic ratios to measure the most fundamental quantity in earth science – absolute geological ages. As stated in the introduction, most of the phenomena that are the subject of earth science studies become perceptible only after the lapse of an extremely long time. An overall understanding of many earth science phenomena can only be obtained by tracing their temporal changes. Consequently, providing an absolute time scale for these phenomena is the most fundamental requirement in earth science research. Up until the 20th century fossils provided virtually the only time scale available for geological phenomena. From a geohistorical viewpoint, however, the appearance of fossils is a comparatively "recent" event that has only occurred over the past one-seventh of the history of the earth. Moreover, fossils merely give relative ages. Thus earth scientists had long dreamt of being able to determine the absolute age of geological phenomena – specifically, that of rocks. The discovery of natural radioactivity in 1896 by A. Henri Becquerel and the subsequent discovery in 1898 of the radioactive element, radium, by M. and P. Curie were the initial steps toward transforming these long-cherished dreams into reality. Focusing on the fact that radioactive decay occurs strictly with a constant rate and that elements such as U and Th have an extremely long half-life rivaling geological time, E. Rutherford suggested in 1905 that these elements might be used as a "clock" to measure absolute geological ages. Numerous subsequent experiments have confirmed that radioactive decay occurs strictly in accordance with the law of radioactive decay:

$$dN/dt = -\lambda N \qquad (4.2)$$
$$T_{1/2} = \ln 2/\lambda.$$

In Eq. (4.2) λ and $T_{1/2}$ stand for the decay constant and the half-life, N is the number of atoms in the radioactive element and t represents time. The decay constant is not affected by the physical environment, such as high temperatures or high pressure, or by the chemical conditions (e.g., the form of the compound) of the material. Therefore the decay constant of radioactive elements contained even in rocks placed in extremely different environments, such as under the high temperatures and high pressure in the earth's interior, or subjected to chemical changes through weathering and so on, is constant, and can be used as an extremely reliable geological clock.

There are many different radiometric dating methods, such as the K–Ar, U–Pb, Rb–Sr, Sm–Nd, Lu–Hf and ^{14}C methods, and these are used for different purposes. Let us discuss the principles and characteristics of several of these dating methods below. Absolute age measurements using isotopes have become increasingly sophisticated in recent years, and are becoming even more of an enigma to those not directly involved in the experiments. Progress in experimental methods has also led to a striking improvement in the precision of experiments, but there are some cases in which greater experimental precision does not necessarily result in a more accurate age scale for geological phenomena. In order to obtain truly meaningful age results from experimental data, correct evaluation of these data is necessary in addition to improved experimental accuracy. Here we will emphasise the evaluation of these data, looking at dating methods using isotopic ratios.

Let us begin by examining the K–Ar dating method, which was the first dating method developed. Naturally occurring K consists of three isotopes, ^{41}K (6.88%), ^{40}K (0.0118%) and ^{39}K (93.10%). ^{40}K is the only isotope that undergoes radioactive decay, disintegrating into two elements, ^{40}Ca (about 90%) and ^{40}Ar (about 10%). Let and stand for the respective decay constants when ^{40}K disintegrates into ^{40}Ar and ^{40}Ca. Here $\lambda = \lambda_e + \lambda_\beta$ is referred to as the total decay constant. $R = \lambda_e/\lambda_\beta$ is called the branching ratio. Let us denote a time axis which measures time t in the past, with the present time being t = 0. This method of denoting the time axis is the method generally used in geochronology. Indicating the present value by the subscript p, the ^{40}K t years ago is $(^{40}K)_p e^{\lambda t}$, and so the K that has decayed over t years is equal to

$$(^{40}K)_p e^{\lambda t} - (^{40}K)_p = (^{40}K)_p (e^{\lambda t} - 1)$$

and this has decayed into ^{40}Ca and ^{40}Ar. Since the amounts of ^{40}Ca and ^{40}Ar are in proportion to the respective decay constants producing them, and λ_e, λ_β,

$$^{40}Ar = \frac{\lambda_e}{\lambda}(^{40}K)_p (e^{\lambda t} - 1).$$

Finding t from this equation,

$$t = \frac{1}{\lambda} \ln \left[1 + \frac{\lambda}{\lambda_e} \frac{^{40}Ar}{^{40}K} \right]. \tag{4.3}$$

Equation (4.3) is the basic equation used in K–Ar age determination.

Let us apply this equation to a specific example. Suppose that volcanic rock erupted and solidified t years ago. At the time of eruption gas components would have escaped from the magma, and so we can assume that ^{40}Ar also escaped. (This supposition is not always strictly correct). Once the rock has solidified, gas can no longer escape, so after solidification the rock preserves all of the ^{40}Ar produced through the decay of the ^{40}K in the rock. The ^{40}Ar in Eq. (4.3) is the amount stored in the rock in this manner. Therefore, using the ^{40}Ar and ^{40}K values found by analyzing the volcanic rock, it is possible to determine from Eq. (4.3) when the rock erupted. This is the principle behind the K–Ar dating method.

The main merit of the K–Ar method is its wide applicability. K is contained in nearly all kinds of rock. Moreover, the half-life of ^{40}K is 1 250 million years, which is shorter than the other long-life nuclei used in dating, such as ^{87}Rb or ^{147}Sm, but long enough, being comparable with the age of the earth. Thus this method can be used to date nearly any geological formation on earth, and it can measure ages from 10^4 to 10^9 years (naturally this depends on the K content). The K–Ar method is virtually the only method for dating basalt younger than about 10 million years old. Thus dating of the oceanic crust, whose main constituent is basalt, is forced to depend completely on the K–Ar method. The disadvantage of this method lies in the fact that since Ar is a gas, it easily escapes from samples. Consequently, this Ar-loss occasionally results in the K–Ar age giving a younger value than the true age. Ar-loss depends on the kind of mineral and the extent of weathering, so selecting samples carefully means that the difficulty of Ar-loss can be avoided to a certain extent. Ar-loss occurs least in amphibole, followed by muscovite and biotite, while most feldspars are poor at retaining Ar, and thus are not suitable samples for K–Ar dating.

When dating samples that are very young or that have an extremely low K content, we must question the assumption that all of the ^{40}Ar in the sample was produced through the decay of ^{40}K. These samples contain extremely little ^{40}Ar produced through the radioactive decay of ^{40}K, so that even if only a tiny amount of inherited ^{40}Ar exists, it can no longer be ignored. In these rocks the ^{40}Ar that is completely unrelated to the decay of the ^{40}K in the rock is known as excess ^{40}Ar. In order to exclude the influence of excess ^{40}Ar and to obtain a meaningful K–Ar age, more complex methods must be adopted, such as the K–Ar isochron method or the ^{40}Ar–^{39}Ar method. Here we will leave the discussion at this point.

The Rb–Sr method uses the radioactive decay of ^{87}Rb into ^{86}Sr ($\lambda = 1.421 \times 10^{-11}$ y, $T_{1/2} = 4.88 \times 10^{10}$ y), and the age t can be found in exactly the same manner as in the K–Ar method by replacing the ^{40}K, ^{40}Ar and λ_e in Eq. (4.3) with ^{87}Rb, ^{87}Sr and λ respectively. In the case of the Rb–Sr method, however, Sr is a solid, not a gas. Therefore, even in cases in which the volcanic rock has solidified from magma, some Sr already exists in the magma, and will be incorporated into the rock. Thus the Sr measured in the rock consists of radiogenic ^{87}Sr formed from the radioactive decay of the ^{87}Rb in the rock after the rock was formed, and ^{87}Sr inherited from at the magma reservoir [known as initial Sr, and written as $(^{87}Sr)_0$]. Thus in order to find the age t, it is necessary to differentiate between the measured ^{87}Sr and that produced through radioactive decay (we will distinguish this by an asterisk). This is done by means of the highly ingenious method known as the isochron method. Specific details of this method were described in Chapter 2 using the dating of meteorites as an example, so we will not repeat the discussion here.

The range of application of the Rb–Sr method is somewhat more limited than that of the K–Ar method. This is because the half-life of ^{87}Rb is almost 50 times as long as that of ^{40}K, and the Rb content in rocks is only hundredths of the K content. Although it lags behind the K–Ar method as far as its range of application is concerned, the Rb–Sr method does have a major advantage not provided by the K–Ar method. Rb–Sr ages are normally found from an isochron. If experimental data are arranged linearly on the $^{87}Sr/^{86}Sr$–$^{87}Rb/^{86}Sr$ diagram, the slope of this straight line will correspond to the age of the sample (see Chap. 2). Suppose now that a material exchange of Rb and Sr has occurred between the sample rock and the external system owing to various geological disturbances. It will not be possible to find an age that is meaningful by applying the Rb–Sr method to such samples. If experiments are actually carried out on samples that have undergone such disturbances and the resulting values are plotted on an isochron diagram, they will not line up on a straight line, that is, they will not form an isochron. In other words, the very linearity of the isochron guarantees the reliability of the ages found. It is as if the dating method itself has a built-in checker that guarantees the significance of the results of the experiment. In order to gain a further understanding of this significance, let us take another look at K–Ar dating. Suppose for now that K–Ar dating is applied to rocks from which some ^{40}Ar has been lost owing to thermal metamorphism and other disturbances. If the ^{40}Ar and ^{40}K are measured, the "age" t can be calculated from Eq. (4.3). However, no matter how precisely ^{40}K and ^{40}Ar are measured, the age t thus calculated is merely a theoretical age value that is irrelevant to the true age of the rock. It is usually difficult to judge whether the age calculated is merely a theoretical age value or whether it represents a truly meaningful rock age. In order to avoid this shortcoming, recently attempts have been made to follow the Rb–Sr method by using an isochron in the K–Ar method also. The ^{40}Ar–^{39}Ar method, which is a variation of the K–Ar isochron method, has also come into widespread use. Both of these methods involve more complex and sophisticated experiments than the simple K–Ar dating method, but the ages obtained can be considered as more meaningful from an earth science viewpoint.

The merits of the Rb–Sr method become even more evident when it is applied to metamorphic rocks. Metamorphic rocks were formed when source rocks such as granite and volcanic rocks underwent metamorphism. Metamorphism is a solid reaction, so that rocks undergoing metamorphism do not melt, and recrystallization occurs inside the constituent minerals as a result of the solid diffusion of ions. Consequently, the crystals comprising the metamorphic rock were formed at the time of metamorphism. If dating using the K–Ar, Rb–Sr or such methods is carried out for individual minerals that have separated from this metamorphic rock, the age found will be the time of metamorphism. Next let us consider the situation when the whole metamorphic rock is used as the sample to be dated. As metamorphism is a solid reaction, the whole rock will not melt, unlike volcanic rocks. The atomic movement at the time of metamorphism is the result of solid diffusion, and this does not go far beyond the range of the crystal size at most unless fluids are present. Thus in the case of the metamorphic rock as a whole, if we take a lump of rock several tens of centimeters large, it can be regarded more or less as a closed system at the time of metamorphism, i.e., the inward and outward movement of atoms can be ignored in comparison with those atoms in the sample. Let us now review the prerequisites for radiometric dating methods. The origin of time when the radioactive decay clock starts ticking is when the system became closed to radioactive elements and the radiogenic elements that are produced through their radioactive decay. As long as the system fulfils the condition of having been closed ever since it began right up to the present, the radio-

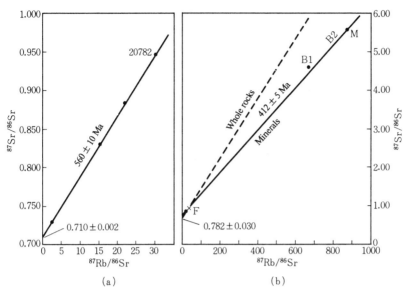

Fig. 4.7. Rb–Sr isochron plots for **a** whole rocks and **b** mineral separates. Samples are from Carn Chuinneag complex, Scotland. A *dotted line* in **b** is the whole rock isochron. Initial $^{87}Sr/^{86}Sr$ ratios are also indicated. (After Long, 1964)

metric age of this sample represents the time from when the system became closed up until the present. Hence if a sufficiently large piece of metamorphic rock (several tens of centimeters in size) is used as the sample, we can expect to find the age of its source rock. There have been numerous reports of excellent Rb–Sr dating of metamorphic rocks, and the age at which the rock underwent metamorphism and the age of its source rock (granite in these examples) have been found separately. As one such instance, Fig. 4.7 shows the classical results for the Rb–Sr method obtained by L.E. Long for a metamorphic rock from Scotland. When the muscovite, biotite and potassium feldspar separated from this metamorphic rock were used to form an Rb–Sr isochron, an age of 412 ± 5 million years was found (shown in Fig. 4.7 by the solid line). The isochron found by using the whole rock as a sample without separating its minerals gives an age of 560 ± 10 million years. This is interpreted as meaning that the granite which was the source rock of this metamorphic rock intruded 560 million years ago, and then later underwent metamorphism 412 million years ago to form the metamorphic rock seen at present (Carn Chuinneag complex). The age found for minerals separated from the rock is known as the mineral age, and that found for the rock itself is known as the whole rock age.

As stated above, the Rb–Sr mineral age and the Rb–Sr whole rock age found for metamorphic rock correspond to the time of metamorphism of the metamorphic rock and the time when its source rock was formed. Unlike the Rb–Sr method, however, when K–Ar dating is applied to mineral samples and whole rock samples it is not possible to differentiate between these two ages. This is because Ar is a gas, and so even though metamorphism is a solid reaction that does not melt the rock, Ar easily escapes owing to thermal disturbances at the time of metamorphism, thus destroying the condition of a closed system. When K–Ar dating is applied to metamorphic rocks, normally the result merely indicates the time of metamorphism, no matter whether a whole rock sample is used or mineral samples are used.

Here we have described how the Rb–Sr method has the outstanding feature of being able to reveal not only the age of metamorphic rocks, but also the time when the source rock was formed, which is beyond the ability of the K–Ar method. This depends on the fact that since the radioactive element Rb and its daughter element Sr are both solid elements, if a large enough rock (of several tens of centimeters) is selected as a sample, it can be regarded as a closed system even at the time of metamorphism. We can expect this characteristic to apply also to other methods using solid element systems, such as the Sm–Nd or Lu–Hf methods. In the Sm–Nd system in particular, both elements belong to the rare earth group, and have very similar chemical properties. Consequently, even if they undergo such external disturbances as metamorphism or weathering, little elemental differentiation occurs, and the Sm/Nd ratio does not easily change at the time of the disturbance. Even supposing that the rock underwent disturbance and the condition of a closed system was violated and some Sm and Nd escaped outside the rock system, as long as the Sm/Nd ratio is kept constant, there is little effect on the Sm–Nd age value. This can be easily understood from the fact that the radiometric age t is given as a ^{143}Nd*/^{147}Sm ratio (the asterisk indicates that it is radiogenic). Hence, some researchers even claim that, depending on the situation, the Sm–Nd clock can go beyond the source rock of metamorphic rocks to the original event, such as the age at which the present mantle separated from the primeval earth. As stated above, both Sm and Nd are rare-earth elements, and their chemical behavior is very similar, so it is difficult for chemical differentiation to occur even if various geological disturbances take place, and the Sm/Nd ratio changes little. Consequently, it is very reasonable to assume that when meteorites and the planets separated from the solar nebula, no major chemical fractionation occurred in the Sm/Nd ratio, and that these bodies all have a common Sm/Nd ratio. Studies of the Sm–Nd system in volcanic rocks and igneous rocks show that rocks older than 2000 million years generally have a value roughly equal to the Sm/Nd ratio of meteorites, but that in rocks younger than about 2000 million years the Sm/Nd ratio gradually deviates from the value in meteorites

(Fig. 3.3b). Many scientists interpret this as evidence that mantle differentiation commenced about 2000 million years ago. This issue has already been discussed in Chapter 3.

As well as enabling us to find the ages of rocks from the slope of the isochron, in such methods as the Rb–Sr and Sm–Nd methods the point at which the isochron intersects the y-axis (see Fig. 2.7) – known as the initial values of the Sr and Nd isotopic ratios, and usually expressed as $(^{87}Sr/^{86}Sr)_0$ and $(^{144}Nd/^{143}Nd)_0$ – also provides us with extremely important information on the origin of rocks. Let us explain this in the case of the Rb–Sr method. As described in Chapter 3, Rb is concentrated in the earth's crust more easily than Sr. Hence the Rb/Sr ratio of the earth's crust is thought to be more than ten times that of the mantle. Consequently, crustal material normally has an $^{87}Sr/^{86}Sr$ ratio that is larger than that of mantle material. The $(^{87}Sr/^{86}Sr)_0$ found from the Rb–Sr isochron is the value inherited from the surrounding material when this rock was formed. Thus the initial Sr isotopic ratio $(^{87}Sr/^{86}Sr)_0$ of the rock and the $(^{87}Sr/^{86}Sr)$ value of the surrounding material where the rock was born should be equal. If the rock was born within the earth's crust, its $(^{87}Sr/^{86}Sr)_0$ value would reflect the high $(^{87}Sr/^{86}Sr)$ value of the earth's crust, while mantle-derived rocks would have a low $(^{87}Sr/^{86}Sr)_0$ value on the whole. Hence the initial Sr value is extremely valuable when dealing with such problems as the origin of material, particularly the question of whether it is of mantle or crustal origin. As one example, Fig. 4.8 shows the $^{87}Sr/^{86}Sr$ values measured for various rocks. Young volcanic rocks that have only recently been derived from the mantle, particularly MORB, which is found where the earth's crust is thin and which has little contamination of crustal material, reflect the $^{87}Sr/^{86}Sr$ value of the mantle, and it is quite obvious that they have a significantly lower $^{87}Sr/^{86}Sr$ ratio than other crustal rocks, which generally have a markedly larger $^{87}Sr/^{86}Sr$ value. Thus the $^{87}Sr/^{86}Sr$ ratio is very effective as a means of discriminating between mantle material and crustal material, and it has played a major role in petrological research.

Since radiometric dating was first established in the 1950's, the subsequent advances in ultra-high vacuum techniques and electronics have led to rapid progress in the accuracy of measurements, and the improved ease and speed of measurements have also been amazing. Radiometric dating is widely used as the most fundamental method in earth science research. Over the past three decades radiometric dating has been carried out so extensively that now there are virtually no major geological structures that have not been dated using this method. Despite these numerous dating results, however, the oldest age for crustal rocks so far is the 3800 million years found for the metamorphic rocks exposed on the west coast of Greenland, reported more than 15 years ago. This is more than 700 million years younger than the age of the earth. At this

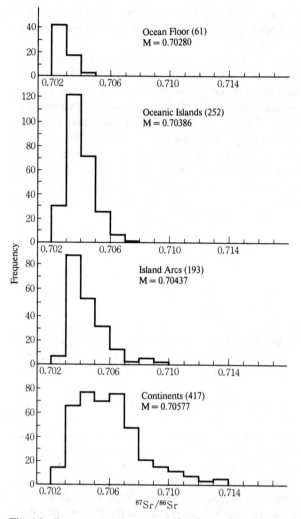

Fig. 4.8. Frequency histograms of $^{87}Sr/^{86}Sr$ ratios of young volcanic rocks in different geological environments. M indicates the arithmetic mean and *numbers* are the number of samples used. (After Faure, 1977)

stage no conclusion can be reached as to whether this indicates that the earth's crust did not exist before 3800 million years ago, or whether it simply means that older rocks do exist somewhere, but have not been discovered yet. Recent geothermal history suggests that the temperature of the primeval earth was close to melting point in the outer part, and it does seem more appropriate to consider that in the early stages of the earth's evolution the crust underwent quite intense thermal disturbances, so that the formation of a stable crust was not possible until more than 700 million years or so after the birth of the earth. A third line of thought is that the meteorite showers that apparently occurred on the moon about 4000 million years ago – even now traces remain in the form of craters on the surface of the moon – occurred on earth in the same period, virtually destroying the earth's oldest crust completely.

Very recently W. Compston and his colleagues applied the U–Pb dating method to zircons from metamorphosed sedimentary rocks collected at Mt. Narryer in Western Australia, and obtained an age of 4100–4200 million years. This may indicate the existence near Mt. Narryer of rocks formed about 4100–4200 million years ago. It will be very interesting to see whether this age is confirmed by the Sm–Nd and Rb–Sr methods.

As far as we can tell from the rocks exposed on the surface of the earth, the oldest earth's crust is about 4000 million years old, considerably younger than the age of the earth itself. A similar conclusion was reached from the discussion of the evolution of the mantle-crust system in Chapter 3. The Nd isotopic ratio data for volcanic and other rocks seem to support the interpretation that the differentiation of the mantle and the earth's crust commenced between 3000 and 4000 million years ago. The question of the exact time of formation of the earth's crust and the question of whether or not the time of its formation differs significantly from region to region remain unsolved, but it seems likely that the earth's crust formed quite some time after the birth of the earth itself.

5 Man and Geohistory

Human beings have existed on earth for over a million years. It is only under extremely limited natural conditions that man and most land animals can survive on earth. Oyxgen, which is essential to the existence of man and other forms of life, accounts for about 21% of the total volume of the present atmosphere. If the amount of oxygen were to fall below 15%, it would become difficult for human beings to maintain life on earth. On the other hand, some scientists have pointed out that if the amount of oxygen were to exceed 30%, great fires would occur on earth. It is only under this delicate balance in nature that the human race can exist. The fact that this balance is not permanent, but has changed continuously throughout geohistory, has been a consistent and repeated theme of this book.

Changes in nature are not confined only to phenomena occurring gradually over a long time in keeping with the evolution of the earth. For example, many geological records indicate that the cataclysmic upheavals in the natural environment that occurred about 65 million years ago at the end of the Cretaceous period destroyed 70% of the life on earth at that time. The theory that this large-scale destruction of life was caused by the impact of bolides (huge meteorites) has been gaining ground in recent years.

External factors such as the impact of bolides are not the only cause of environmental changes that may threaten the very foundation of existence of the human race and other forms of life. The tremendous consumption of fossil fuels as an energy source in the 20th century, particularly in the latter half, has shown that the existence and activities of man have a profound effect on how the earth will evolve in the future. Worldwide consumption of energy in 1980 has been estimated at approximately 5×10^{12} Watt. This colossal amount is equivalent to as much as one twenty-thousandth of the total energy reaching the earth from the sun (approximately 10^{17} Watt per total cross-section of the earth). Until the massive consumption of fossil fuels by man commenced, the earth had continued to evolve over 4500 million years in complete disregard of the existence and activities of human beings. In the latter half of the 20th century, however, the energy produced by man has reached a scale where it cannot be ignored even in comparison with solar energy, and the future evolution of the earth can no longer escape the effect of human activities. The increased CO_2 concentration in the atmosphere has already appeared as a clear global trend as the natural outcome of the enormous consumption of fossil fuels. It has been forecast that this increase will also have a serious effect on climatic variations.

In this chapter we will discuss several events in geohistory, such as meteorite impacts on the earth and a naturally occurring uranium chain reaction. These events may appear to be rather unusual, but we will see in this chapter that they are intimately related to such problems as global climatic changes due to an increase of CO_2 in the air or to the disposal of lethal radioactive wastes. These may result in a fatal disaster in the foreseeable future unless quick and effective measures are taken. We will also show that studies on these past events provide very useful clues to approaching these formidable problems confronting the world today. This is another use of geohistory besides its purely academic interest.

5.1 Bolide Impacts: Mass Extinction of Life?

a) K–T Boundary

The boundary (known as the K–T boundary) between the Mesozoic era (240–65 million years ago) and the Cenozoic era (from 65 million years ago up to the present) is characterized by a large-scale mass extinction of life. It has been inferred that at the end of the Cretaceous period (65 million years ago), which is the borderline between the Mesozoic and Cenozoic eras, more than 70 percent of the living species in existence at that time suddenly disappeared. In the Cretaceous period the great reptiles that were in their heyday in the Mesozoic era, and their enormous symbol, dinosaurs, also disappeared completely from the stage of the earth's evolution. This mass extinction of life was not confined to large land creatures, such as dinosaurs and flying reptiles. Anmonites and bivalves, and even zooplankton and photo plankton, did not escape annihilation. What led to this large-scale mass extinction at the end of the Cretaceous period? A decade ago, L.W. Alvarez and his colleagues sparked off new controversy over this issue that has long evoked great argument amongst geologists by their discovery of an abnormally high Ir content in the Cretaceous-Tertiary boundary of sedimentary rocks near Gubbio in northern Italy.

Along with platinum and osmium, Ir belongs to the platinum group of elements. As with platinum, very little Ir exists in the earth's crust (approximately 0.1 ppb), so it is indeed worthy of the name "rare metal". Meteorites, however, contain as much as 1000 times the amount found in crustal rocks. Since Ir is a nonvolatile element, the earth as a whole should contain about the same amount as in meteorites (see Chap. 3.2). The extremely small amount in crustal rocks compared to meteorites is explained that because of the siderophile nature, i.e., high affinity with iron, most of terrestrial Ir as well as other siderophile elements reside now in the metallic core. Alvarez and his colleagues took samples from a sedimentary stratum near Gubbio in Italy that is regarded as a faithful record of the Cretaceous-Tertiary boundary stratum, and analyzed the Ir values of these samples. They discovered an anomalous concentration of Ir in this stratum. The Ir content, which is less than 1 ppb below the boundary stratum, reaches several tens of ppb in the sector corresponding to the boundary stratum.

Prior to this discovery by Alvarez and his colleagues, some researchers had already proposed that the large-scale mass extinction of life that characterizes the boundary between the Cretaceous and Tertiary periods was caused by extraterrestrial objects, such as meteorites, colliding with the earth. However, there was no concrete evidence in support of this hypothesis. The discovery by Alvarez and his colleagues of the anomalous Ir concentration in the Cretaceous-Tertiary boundary immediately transformed this hypothesis into a highly realistic and convincing argument. Since then anomalous Ir concentrations have also been discovered in sediment in Denmark and Spain and in the Pacific and Atlantic Oceans, and it has become recognized as a global phenomenon. Concentrations of Co, Os, and Pt, which are characteristically concentrated in extraterrestrial material, have also been reported in this boundary stratum. As one example we have shown in Fig. 5.1 the results obtained for

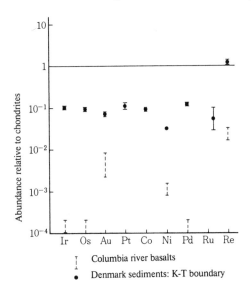

Fig. 5.1. Noble metal concentrations in the Cretaceous-Tertiary boundary clay from Denmark (*solid circles*) and in Columbia River basalts (*dotted lines*). The concentrations are normalized to C1 chondrites. (After Ganapathy, 1980)

a K–T boundary stratum in Denmark that has been studied by R. Ganapathy. As Fig. 5.1 reveals, it is obvious that such elements as Ir and Os are concentrated here ten to a hundred times more than in crustal material (basalts from the Columbia River in America were used in this comparison). The relative abundances among the heavy metals in this sediment are also very similar to those in C1 chondrites. The elements in Fig. 5.1 are all heavy metals that are markedly more concentrated in meteorites than in crustal material. Supposing that this concentration of heavy metals in the Cretaceous-Tertiary boundary is the result of the inclusion of fragments of C1 chondrites, then their proportion would be about 7 to 8 percent. Let us also suppose that the anomalous concentration of heavy metals in the sedimentary stratum in Denmark is common throughout the world. This degree of heavy metal concentration in the 2-centimeter thick Cretaceous-Tertiary boundary all over the world can be explained by assuming that a meteorite with a diameter of about 11 km impacted with the earth, where it was smashed to pieces and scattered uniformly.

The isotopic ratios of rare gases have also been measured in an attempt to find direct proof of the inclusion of extraterrestrial material in the Cretaceous-Tertiary stratum, but so far there have been no reported cases of these isotopic ratios showing a value characteristic of extraterrestrial material. On the other hand, microtektites have been discovered on a global scale in the deep sea sediment Cretaceous-Tertiary boundary that has an anomalous concentration of Ir. A tektite is a kind of glass that has a peculiar button-like shape. It is found at special locations on earth, and is thought to have been formed when the impact of falling meteorites melted crustal rocks locally and scattered them. The world-wide existence of microtektites in the Cretaceous-Tertiary boundary supports the Alvarez hypothesis that bolides impacted with the earth and their fragments scattered around the whole world.

Let us suppose that the anomalous concentration of heavy metals in the Cretaceous-Tertiary boundary first reported by Alvarez and his colleagues and subsequently confirmed by many researchers is of extraterrestrial origin, and that it is the result of the impact of bolides. If a meteorite about 10 km in size collided with the earth, the energy produced by this impact would be enormous, amounting to 3×10^{24} erg, and the effect on the earth defies imagination. Alvarez and his colleagues claimed that the impact of bolides was the direct cause of the annihilation of life at the end of the Cretaceous period. The mechanism by which this impact actually led to the large-scale extinction of living species is still a matter of conjecture. One theory attributes this to the collapse of the food chain system, suggesting that when the meteorite fell in the sea it caused enormous tsunami that annihilated marine plankton, and as a result the fish that had depended on plankton for food became extinct, and this eventually affected life on land. Another hypothesis is that the dust raised at the time of the bolide impact covered the earth and screened solar radiation, leading to a sudden temperature drop that killed most animals. Many other theories have been proposed, but none could be described as a final conclusion.

b) The Collision of the Earth and Meteorites

The anomalous condensation of heavy metals in the Cretaceous-Tertiary boundary strongly suggests a large-scale inclusion of extraterrestrial material. It is not difficult to imagine the enormous influence of this on the ecosystem, and this seems to substantiate the theory attributing the extinction of life at the end of the Cretaceous period to the impact of bolides. On the other hand, is such an impact of bolides (with a diameter of about 10 km) actually possible from an astrophysical viewpoint? In the solar system countless numbers of objects besides the planets are moving around the sun in a Kepler motion, including asteroids and comets. Some of these have been observed to have a Keplerian trajectory that intersects the earth's orbit (the Keplerian motion of the earth around the sun). The possibility does exist that several of these objects may actually collide with the earth, and the probability of such a collision can be estimated theoretically. Figure 5.2 shows an example of such an estimate by E.M. Shoemaker. The horizontal axis denotes the energy of the object colliding with the earth, and the vertical axis denotes the cumulative frequency of the collisions. Here the impact energy (W) is given as

$$W = (1/2) M v^2 /(4.18 \times 10^{10}) \; g \cdot TNT \tag{5.1}$$

calculated in terms of the explosive energy of TNT (unit: gram). M is the mass (g) of the object colliding with the earth and is the velocity (cm s^{-1}), and the denominator corresponds to the amount of energy released when one gram of explosives explodes (1 g TNT $\cong 10^3$ cal = 4.18×10^{10} erg). In Fig. 5.2 the fact that the collision frequency declines rapidly as the energy of the colliding object increases reflects the size distribution of objects in the solar system – the larger the object, the more rapidly their number declines. Figure 5.2 was drawn up on the basis of observations of comets and asteroids and on theoretical calculations of their trajectories. How frequently do these bodies actually collide with the earth? The recent progress in planetology has led to attempts to examine the traces on earth of its collisions with meteorites and other extraterrestrial bodies. A recent example of such meteorite craters is that caused by a meteorite that fell near Reverstoke in British

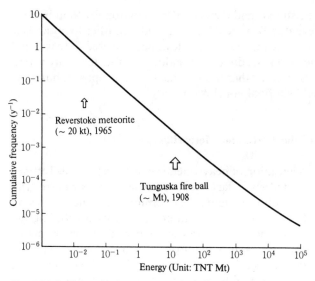

Fig. 5.2. Collision frequency (cumulative) of meteorites onto the earth versus the impact energies (in mega-tons of TNT). *Arrows* show Reverstoke meteorite and Tunguska fire ball impact events. (After Shoemaker, 1983)

Columbia, Canada on March 31, 1965. Apparently the sky lit up even 800 km away from the site where the meteorite fell. From the intensity of the shock waves produced when the meteorite entered the atmosphere, it has been estimated that the impact energy was equivalent to about 20 000 tons of TNT. If the meteorite fell at a rate of 20 km s^{-1}, its weight can be calculated from Eq. (5.1) as approximately 500 tons. Small fragments of a carbonaceous chondrite and countless pieces of round glass were found on the snow near Reverstoke. The pieces of glass are earth rocks that melted in the heat of the meteorite's impact and were scattered around. On an even larger scale is the famous "Tunguska fireball". On June 30, 1908 a "fireball" thought to be caused by a meteorite that fell near Tunguska in central Siberia mowed down more than 40 km of forest along the direction of the meteorite's fall. The light was seen as far away as England, and letters written to British newspapers at the time stated that it was bright enough to read a newspaper in the middle of the night. The noise of this fireball's impact was heard more than 1000 km away, and from analyses of the seismic waves at that time A. Ben-Menaham has estimated its impact energy at 12.5 ± 2.5 megatons (TNT equivalent). Despite this terrific energy, however, curiously enough absolutely no meteorite material has been discovered on the ground in the Tunguska region so far. All that has been found are microscopic glass balls formed by rocks (of earth origin) melting. Hence it has been concluded that the Tunguska fireball exploded at quite a high altitude (up to 8.5 km high) by the impact of its entry into the atmosphere, and that it scattered without reaching the earth intact. The fireball probably consisted mainly of comparatively fragile rocks or the ice seen in comets. Figure 5.2 shows the examples of Reverstok and Tunguska. From the fall-frequency estimate curve it can be inferred that the events at Reverstok and Tunguska are of a type that occur once every several years and once every several hundred years respectively. Looked at in this light, we can understand that even events that have wreaked great destruction on the surface of the earth, such as at Tunguska, are by no means a rare phenomenon from the viewpoint of geohistorical time. Going further back into the past, a meteorite crater in Arizona that is the result of an iron meteorite that fell about 25 000 years ago remains in almost perfect shape. The diameter of the crater is about 1.2 km, and it is presumed to have been formed as the result of about 15 megatons of impact energy. Meteorites are likely to have fallen in geological times with about the same frequency as today, but owing to erosion on the surface of the earth it becomes more difficult to confirm the existence of older craters that would provide direct proof of the impact of such meteorites. However, there have been reports around the world of cases in which it is assumed from the surrounding geological structures that huge meteorites have fallen. The energy of the bolide impact that is concluded to have occurred at the boundary between the Cretaceous and Tertiary periods is approximately 10^3 megatons TNT, and if the frequency curve in Fig. 5.2 is applicable, we can infer that events on this scale occur every 100 000 years or so. If we consider a period of 100 million years or so, it is quite conceivable that bolides of this size will collide with the earth, at least as far as we can tell from Fig. 5.2.

The theory that the impact of such bolides led to the mass extinction of life that characterizes the Cretaceous-Tertiary boundary originated in the discovery by Alvarez and his colleagues of an anomalous concentration of Ir in the Cretaceous-Tertiary boundary stratum. Even today this is regarded as one of the strongest pieces of evidence supporting this theory. However, is the inclusion of meteorite material the only cause of this Ir concentration? As stated in the previous section, the fact that the Ir content is lower than in meteorites is merely a characteristic of crustal material that has differentiated from the mantle and the core. From the meteorite analogy (see Chap. 3.2), we have assumed that the earth and meteorites contain a similar amount of nonvolatile elements such as Ir. Thus we cannot immediately conclude that the anomalous Ir concentration is the result of the inclusion of meteorite material. In fact, anomalous Ir concentrations have also been found in sedimentary strata other than the Cretaceous-Tertiary boundary. Moreover, D.V. Kent, as well as C.B. Officer and C.L. Drake and their colleagues, has cast strong doubts on the bolide theory. They have pointed out that although, judging from the size of its caldera, the effect on the atmosphere of the eruption of the Toba volcano 75 000 years ago rivaled that of the alleged bolide impact on the Cretaceous-Tertiary boundary, it did not lead to large-scale mass extinction. Officer and Drake have suggested that the world-wide anomalous concentration of heavy metals such as Ir and Co in the Cretaceous-Tertiary boundary may be the result of extremely active volcanic activity at that time. According to their explanation, this volcanic activity resulted in Ir and other heavy metals being brought up from the mantle and then scattered throughout the world together with volcanic ash. As far as the mass extinction is concerned, they suggest that even if a large-scale eruption of a volcano like the Toba volcano occurred, it would only have a temporary effect on living species, and would not lead to a large-scale annihilation of bio-mass. If several such eruptions were to occur in succession, however, they would have a tremendous effect on the atmosphere, and it would not be surprising if a large-scale mass extinction of life were to occur.

c) Volcanic Ash and the Fall in Temperature

The discovery of the anomalous Ir concentration in the sedimentary stratum in Northern Italy nearly a decade ago gave new life to the theory that the large-scale mass extinction of life in the Cretaceous-Tertiary boundary was caused by the impact of bolides. However, there is also a strong claim that inclusion of extraterrestrial material is not the sole cause of the Ir concentration, but that it can be also explained by supposing that Ir and other rare metals were derived from the mantle in the wake of volcanic activity. Both of these theories, however, assume that the events that sprinkled Ir and other heavy metals around the world – large-scale volcanic eruptions or the impact of bolides – raised large quantities of dust in the atmosphere, thus screening the solar radiation and causing the temperature to drop suddenly, eventually leading to massive mass extinction. The answer to the question of whether the mass extinction of life at the end of the Cretaceous period was the result of the impact of bolides or of intense volcanic activity must await further research. Meanwhile, the bolide theory that originated in the anomalous Ir concentration has subsequently developed in an unexpected direction. This is the issue of the relation between the increased amount of dust in the atmosphere and changes in the temperature.

The bolide that is regarded as the main cause behind the extinction of many biological species at the end of the Cretaceous period is assumed to have been about 10 km in diameter. The energy released upon colliding with the earth would be about 3×10^{24} erg, equivalent to the energy of several hundred megaton-class hydrogen bombs. If several hundred nuclear bombs were to explode at once, they too would scatter large quantities of ash and dust. Is it not possible that this dust would block out the sunlight and that the temperature would drop sharply, bringing on a "nuclear winter" and eventually leading to a widespread annihilation of life rivalling that in the Cretaceous period? This question sparked off the nuclear winter debate that has been attracting interest among many scientists recently. Thus the Cretaceous-Tertiary boundary issue, symbolized by the extinction of dinosaurs, has a direct bearing on very realistic issues today. Meteorologists are divided as to whether or not a nuclear winter will actually occur. Dust would not be the only effect of a bolide impact or a nuclear explosion on the climate. It is predicted that, along with volcanic ash and dust, the amount of H_2O and CO_2 in the atmosphere would also increase enormously. The sulfur and nitrogen oxides formed in the upper atmosphere would also be important in climatic changes. Since CO_2 and H_2O absorb the infrared radiation from earth, this would cause the atmospheric temperature to rise. The reduction in sunlight resulting from the volcanic ash and dust would be partially offset by this temperature increase. Many other phenomena are also intertwined in an extremely complex fashion, and it is very difficult to estimate correctly the long-term effect on the earth of volcanic eruptions or nuclear explosions. This discussion is beyond the scope of this book, so here we will confine ourselves to a geohistorical approach to this issue.

By examining past records we can estimate to a certain extent the effect of large-scale volcanic activities on climatic variations. The recent eruption of the El Chichon volcano in southern Mexico on April 4, 1982 is expected to provide excellent data for investigating eruptions and their effect on the climate. Although the scale of the eruption was smaller than the St. Helens eruption 2 years earlier, El Chichon erupted in a more or less vertical direcion, so it is estimated to have spewed forth six to seven times the amount of material emitted by St. Helens – nearly 500 million tons in all. The volcanic plume reached a height of 35 000 m, and even six months after the eruption volcanic ash was clearly observed floating in the air over Hawaii Island nearly 10 000 km away from the volcano. As of 1985 no clear answer has been obtained to the question of how the El Chichon eruption has affected the temperature. Doubtless, however, extremely valuable conclusions can be obtained if combined with data on global climatic variations in the future. The largest volcanic eruption of which historical records remain is the eruption of the Tambora volcano in Indonesia on April 5, 1815. It is estimated that the volcanic ash and dust emitted into the atmosphere was more than 100 times the amount discharged by St. Helens. After the eruption of Mt. Tambora, Europe and North America suffered record cool summers in 1816 and

1817 with summer frosts leading to catastrophic crop failures. However, we cannot immediately attribute these cool summers to the eruption of Mt. Tambora. There have been many cool summers in history which cannot be linked to a volcanic eruption. The cool summers in 1816 and 1817 may merely have been an example of such climatic variations. Statistical simulation is essential to throw light on this question. Taking the effect of volcanic dust into account and using data (estimates of the amount of volcanic dust) on volcanic eruptions for which records remain, S.H. Schneider and C. Mass carried out model calculations of climatic variations, and reached the conclusion that volcanic dust has a significant effect on climatic variations. Recent results of similar numerical simulation calculations have led to the conclusion that the average temperature around the world fell by almost 1 °C for a year or two after the Tambora eruption. If several eruptions on this scale occurred in succession over a short period of time, or if nuclear explosions occurred on a global scale, the effect on the climate would be far more serious.

5.2 The Fate of Radioactive Waste – The Oklo Phenomenon

The massive consumption of oil that commenced in the latter half of the 20th century has greatly boosted the concentration of CO_2 in the atmosphere. Scientists are concerned that the increased CO_2 concentration may prevent heat radiation from the earth and eventually lead to a significant rise in the average temperature. The issue of an increase in the CO_2 concentration is highly symbolic of how the course of geohistory, which up until now has continued its evolution in complete disregard of the existence of man, may be governed in the future by the activities of man. Some activities of human beings, although perhaps not directly affecting the earth's evolution, are reaching such a scale and becoming so serious that they may shake the very foundation on which the existence of man stands. One of the most urgent and serious problems facing man today is the issue of the radioactive waste produced by nuclear reactors. Some nuclear waste has harmful radiation that lasts over a very long time, so measures dealing with this must cover an extremely lengthy period of 10^4 to 10^5 years, or even longer. Measures on a geohistorical time scale are called for here. Man's involvement with radioactive material commenced in 1896 with the discovery by H. Becquerel of the radioactivity of uranium ore, so our experience in this field goes back less than a century. On the other hand, the radioactive waste problem requires countermeasures on a time scale exceeding the entire history of the human race. The geohistorical approach provides very effective clues to a solution of this problem. It is less than 50 years since the first nuclear reactor was built by man, but a natural nuclear reactor was already in operation 2000 million years ago. This is the Oklo natural reactor discovered in central Africa in 1972. Here it is possible even today to trace a past nuclear chain reaction involving uranium, that is, the vestiges of the operation of a nuclear reactor. The remains of the Oklo nuclear reactor will provide an inkling of the future of the nuclear reactors made by man and the waste produced by these. This knowledge will offer incomparable clues on how to solve the problem of the disposal of nuclear waste.

a) The Oklo Natural Nuclear Reactor

On September 25, 1972 the French Atomic Energy Commission announced that uranium quarried from the Oklo uranium mine in the eastern part of the Republic of Gabon near Gabon's border with the Congo had a highly anomalous isotopic ratio ($^{238}U/^{235}U$). It announced that the reason for this was a nuclear chain reaction involving uranium that had occurred in the past at Oklo. This discovery provided proof that the uranium chain reaction carried out in 1942 for the first time in man's history by E. Fermi and his colleagues at the University of Chicago had actually occurred in nature 2000 million years ago. The chain reaction involving uranium constitutes the very heart of nuclear reactors today, and it is no exaggeration to describe the Oklo chain reaction as a natural nuclear reactor.

What sparked off the discovery of the Oklo natural nuclear reactor was the anomalous isotopic composition of its uranium. Naturally occurring uranium consists of three isotopes, ^{234}U, ^{235}U, and ^{238}U. As discussed in Chapter 2, with a few exceptions the isotopic compositions of elements constituting solar material are highly uniform. Uranium is no exception. As far as measurements so far show, ^{234}U, ^{235}U, and ^{238}U exist at present in meteorites and earth material in the respective ratios of 0.0054, 0.72, and 99.275% (number of atoms), and within the range of experimental error (approximately 0.1%) they show virtually identical values. In surprising contrast, however, the proportion of ^{235}U in the uranium from the Oklo mine was found to vary between 0.72 and 0.33%. Naturally, even isotopic ratios can vary to a certain extent as the result of mass fractionation under the conditions in nature. However, it is impossible for the isotopic ratio of an element with a large mass number, like uranium, to more than double simply due to mass fractionation. To explain this drastic variation in the isotopic ratio, we must consider some form of nuclear reaction. As will be discussed later, in addition to uranium, other rocks collected from the Oklo mine have also demonstrated the isotopic variations characteristic of a uranium fission, such as in certain kinds of rare earth elements and rare gases, thus demonstrating conclusively that a nuclear fission occurred in the past in the Oklo mine area. Moreover, the concentration of uranium and the anomalous isotopic ratios of the rare earth elements correlate extremely well, indicating that the direct cause of the anomalous isotopic ratio lies in a nuclear reaction involving uranium.

The possibility of a fission of naturally occurring uranium had already been discussed quite seriously among many scientists at the time of the successful nuclear fission experiment by Fermi and his colleagues. But they concluded that a nuclear fission by uranium could not occur under natural conditions. For the fission to occur, it is necessary for the speed of the neutrons emitted from ^{235}U to be reduced, and so a sufficient quantity of a moderator – H_2O is the most efficient moderator in nature – must exist right beside the uranium atoms. Scientists of the time concluded that this condition is not fulfilled in nature. P.K. Kuroda took up this issue again in 1954. Focusing on the fact that when uranium is deposited in veins it crystallizes from a hydrothermal solution, Kuroda pointed out that a sufficient quantity of water to reduce the speed of the neutrons always exists around uranium ore after its crystallization. Another factor conducive to a fission of natural uranium is the effect of time. Of the three isotopes constituting natural uranium, ^{235}U plays the main role in nuclear chain reactions. ^{235}U has a half-life of 7.04×10^8 years. Consequently, the relative concentration of ^{235}U about 700 million years ago was almost double the present amount, and nearly eight times the current amount existed in the Precambrian period 2000 million years ago. Considering these two factors, Kuroda pointed out that in the Precambrian period natural uranium would have been fully capable of producing a nuclear chain reaction under natural conditions. However, Kuroda's claim was completely ignored until the existence of a "natural nuclear reactor" was verified for the first time at Oklo. Recalling those days, in a later paper Kuroda quoted with some irony the words of a leading scientist of the time: "Some of the world's best physicists had constructed the Stagg Field reactor, the world's first reactor, constructed beneath the Stagg Field athletic field at the University of Chicago with careful attention to mechanical detail, to the purity of materials and to the geometry of the assembly. Could nature have achieved the same result so casually?"

Conclusive proof that a nuclear chain reaction involving uranium occurred in the Oklo mine area was provided by the anomalous isotopic ratios of uranium and the rare earth elements there. When uranium undergoes a nuclear fission, ^{235}U is consumed in the nuclear fission, while such isotopes as Nd, Sm, and Gd are produced as fission products. Consequently, in rocks that have undergone nuclear fission of uranium

Table 5.1. Nd isotopic compositions

	^{142}Nd	^{143}Nd	^{144}Nd	^{145}Nd	^{146}Nd	^{148}Nd	^{150}Nd
Oklo	1.38	22.1	32.0	17.5	15.6	8.01	3.40
Oklo	5.49	23.0	28.2	16.3	15.4	7.70	3.90
Fissiogenic Nd	0	28.8	26.5	18.9	14.4	8.26	3.12
Natural Nd	27.11	12.17	23.85	8.30	17.22	5.73	5.62

the relative abundance of ^{235}U is less than in ordinary rocks, while the abundance of some isotopes, such as Nd, Sm, and Gd isotopes, will increase. Table 5.1 shows the isotopic composition of neodymium (Nd) found for uranium ore quarried at the Oklo mine. For the sake of comparison, the table also shows the isotopic composition of ordinary natural neodymium from other sites. This comparison reveals that the Oklo mine rocks have a markedly different neodymium isotopic composition from that of ordinary neodymium. Table 5.1 also shows the isotopic composition of neodymium formed through the nuclear fission of ^{235}U. Note that the isotopic composition of the neodymium from Oklo mine resembles that of the fission product of ^{235}U more closely than it resembles natural neodymium. The reason that they do not match perfectly is that the neodymium from the Oklo mine contains some natural neodymium in addition to the neodymium produced through the nuclear fission of uranium. If a correction is made for the latter, the neodymium from the Oklo mine has the same isotopic composition as neodymium produced through nuclear fission from ^{235}U.

There is no room for doubt about the fact that the anomalous isotopic compositions of the uranium and neodymium found at the Oklo mine were produced through a nuclear fission involving uranium. When did this nuclear reaction occur, and on what scale? Let us begin by considering the first problem. The uranium ore at Oklo mine is quarried from Precambrian clay sandstone. Figure 5.3 shows an outline of the geological structure of the Oklo natural nuclear reactor. The area with the anomalous uranium and neodymium isotopic compositions is known as the nuclear reactor zone, and is 15 m long and less than 1 m in width, and is covered in this clay sandstone. The uranium concentration in the nuclear reactor zone is very high, reaching 30 to 50%. The nuclear reactor zone corresponds to the core of a man-made nuclear reactor, and seems to have originally consisted of uraninite that was enriched from clay sandstone similar to the surrounding area. The K–Ar ages found for this clay sandstone are quite scattered around 1800 million years. It is by no means clear how the K–Ar ages found for the sedimentary rock are related to the time of sedimentation and the formation age of the uranium deposits. Even if we determined the time of sedimentation, the time of the commencement of the chain reaction at

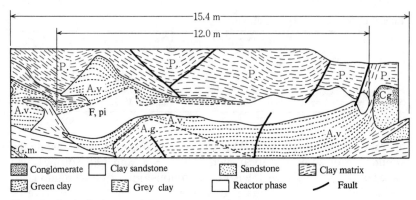

Fig. 5.3. Geological map of Oklo natural reactor. (After Gauthier-Lafaye, 1978)

the Oklo natural nuclear reactor cannot be linked directly to the age of the sedimentary rock. As shown below, this time can be deduced independently from the variations in the uranium isotopic ratio.

Let us first consider the temporal change in ^{235}U when uranium sets off a chain reaction. Here we will consider that the duration of the chain reaction was so much shorter than the half-life of ^{235}U (7.04×10^8 years) that it can be ignored, and we will not take the effect of the radioactive decay of ^{235}U into account. Hence the time variation in ^{235}U is

$$\frac{d}{dt}{}^{235}U = -\sigma\phi\,{}^{235}U + C\sigma\phi \cdot {}^{235}U. \tag{5.2}$$

Here σ indicates the total neutron capture cross-section of ^{235}U, and ϕ indicates the neutron flux intensity. The first term shows the fission of ^{235}U by neutrons, and the second term shows the production of ^{235}U resulting from the neutron capture reaction of ^{238}U. C is called the reproduction coefficient, and is a constant characteristic to each different nuclear reactor. Specifically, it shows the ratio between the ^{235}U that undergoes fission in the chain reaction and the ^{235}U that is reproduced. Solving Eq. (5.2) gives

$$^{235}U_t = {}^{235}U_0 \cdot \exp[-(1-C)\sigma \cdot \phi \cdot t]. \tag{5.3}$$

Here $^{235}U_0$ is the initial value immediately before the chain reaction commences, and $^{235}U_t$ shows the value when the chain reaction terminates. t is the duration of the chain reaction. Supposing that the chain reaction finished T years ago, $^{235}U_t$ will undergo radioactive decay up to the present and decrease, so that the present value of ^{235}U will be

$$^{235}U_p = {}^{235}U_t\, e^{-\lambda_{235} \cdot T}, \tag{5.4}$$

where λ_{235} denotes the decay constant of ^{235}U.

From Eqs. (5.2) and (5.4)

$$^{235}U_p = {}^{235}U_0 \cdot \exp[-(1-C)\sigma\phi \cdot t]e^{-\lambda_{235}T}. \tag{5.5}$$

The number of nuclear fissions in the unit time as the result of ^{235}U reacting with the neutrons is

$$\frac{d}{dt}{}^{235}U_f = \sigma_f \cdot \phi \cdot {}^{235}U. \tag{5.6}$$

Here σ_f is the nuclear fission cross section of ^{235}U [note that the σ in Eq. (5.2) is the total cross-section for all neutron reactions, and differs from σ_f]. Thus from Eqs. (5.3) and (5.6)

$$^{235}U_f = {}^{25}U_0 \cdot \frac{\sigma_f}{(1-C)\sigma} \cdot [1 - \exp\{-(1-C)\sigma\phi \cdot t\}]. \tag{5.7}$$

In order to find t, Eq. (5.7) is divided by $^{238}U_p$:

$$^{235}U_f/^{238}U_p = {}^{235}U_0/^{238}U_p \cdot \frac{\sigma_f}{(1-C)\sigma} \cdot [1 - \exp\{-(1-C)\sigma\phi t\}]. \tag{5.8}$$

Taking into consideration the fact that the isotopic ratio (present value) of natural uranium that has not been involved in a chain reaction is constant ($^{235}U/^{238}U = 137.8$),

$$^{235}U_0/^{238}U_P = (^{235}U/^{238}U)_P \cdot e^{\lambda_{235} T}.$$

So Eq. (5.8) immediately produces

$$^{235}U_f/^{238}U_p = \frac{1}{137.8} \cdot e^{\lambda_{235} T} \cdot \frac{\sigma_f}{(1-C)\sigma} \cdot [1 - \exp\{-(1-C)\sigma\phi t\}]. \tag{5.9}$$

The left-hand side in Eq. (5.9) is the number of ^{235}U nuclear fissions per ^{238}U nucleus, and can be estimated from the deviations in the isotopic compositions of rare earth elements (Nd, Sm) at the same site. $\phi \cdot t$ are total neutron fluxes that were involved in the chain reaction, and can be deduced from measurements of the extent to which ^{235}U has decreased compared with the isotopic composition of ordinary uranium (that has not undergone a chain reaction). C andare constants, and can be estimated empirically. Substituting these figures in Eq. (5.9) gives the value of $T \cong 2 \times 10^9$ years as the time since the chain reaction finished, that is, the age at which the chain reaction occurred. This value is in line with the geological results discussed earlier. Moreover, the duration of this chain reaction can be estimated from the ratio of nuclear fissions of ^{238}U and ^{239}Pu (formed in the process of the chain reaction) in the chain reaction. It has been estimated that the chain reaction lasted approximately 500 000 years.

As shown in Fig. 5.3, the "core" region at Oklo mine currently extends over a length of about 10 m. The scale of the chain reaction that occurred at the Oklo natural nuclear reactor 235 can be inferred from the amount of ^{235}U that "burnt" as fuel. In extreme cases, the ratio of ^{235}U in the uranium ore at Oklo mine has decreased to nearly half the amount in ordinary natural uranium. Obviously, this was burnt up as nuclear fuel. It is estimated that the total amount of ^{235}U consumed as nuclear fuel in the whole Oklo nuclear reactor region was approximately six tons. Since the energy released through the nuclear fission of 1g of ^{235}U is approximately 10^{11} Joules, six tons of ^{235}U would release 6×10^{17} Joules of energy. As stated above, the "operating period" of the Oklo natural nuclear reactor is estimated to have been about 500 000 years, so the power produced was a mere 30 kW or so. Hence the Oklo natural nuclear reactor was on an extremely small scale compared to the commercial nuclear reactors in operation today, which produce millions of kilowatts of power.

b) The Environment and Radioactive Waste

Two thousand million years ago nature had already produced a "nuclear reactor", though it is only recently that man has succeeded in this. Until Fermi and his colleagues succeeded with their experiments in 1942, man was completely ignorant of nuclear chain reactions and the various problems entailed. Nuclear chain reactions are having an incalculable impact on all aspects of human society, for both good and bad. On the negative side, the issue of disposing of the rapidly accumulating radioactive waste is particularly serious. The difficulty of this problem lies in the fact that, while urgent measures are called for, the disposal of this waste must be considered on a geohistorical time scale that also takes into consideration radioactive material with a long half-life of tens or hundreds of thousands of years. Whether radioactive material is abandoned or whether it is stored, we have no means of judging the safety of these methods in the empirical time of our laboratories. How can we understand phenomena that occur on a long time scale far exceeding our experience? The Oklo natural nuclear reactor is an excellent laboratory for solving this problem.

The total output of the Oklo natural nuclear reactor matches that of five 100 million kW commercial nuclear reactors operated for one year. The amount of radioactive material produced is on the same scale. The radioactive material produced by the Oklo natural nuclear reactor has been left undisturbed for nearly 2000 million years. Thus an examination of the Oklo mine will reveal how this radioactive material has affected the surrounding environment over this long time span.

Plutonium is one of the most dangerous and difficult-to-handle radioactive waste products of commercial nuclear reactors today. Several hundreds of kilograms of Pu are produced annually as a by-product of the operation of large nuclear reactors. ^{239}Pu, is a highly toxic isotope that undergoes α disintegration with a half-life of 24 000 years. ^{239}Pu was also produced in the Oklo natural nuclear reactor, and the amount has been estimated at about three tons. Let us now consider the problem of how this ^{239}Pu, which had been left in its natural state, behaved until it had completely decayed after hundreds of thousands of years.

The isotopic ratios of uranium collected from the Oklo mine vary significantly, but are characterized by the fact that they always show a ^{235}U deficiency. This ^{235}U deficiency is because it was consumed as the nuclear fuel for the natural nuclear reactor. On the other hand, ^{239}Pu is formed through the reaction with neutrons

$$^{238}U (n, \gamma) \,^{239}U \rightarrow \,^{239}Np \rightarrow \,^{239}Pu \rightarrow \,^{235}U \qquad (5.10)$$

22.5 min 2.35 days 24 100 years,

and this ^{239}Pu decays to produce ^{235}U. In the reaction in Eq. (5.2), while one atom of ^{235}U is consumed through nuclear fission, approximately 0.45 atoms of ^{239}Pu and the same number of radiogenic ^{235}U atoms are formed, as was described earlier [see Eq. (5.2)]. Suppose that ^{239}Pu separated from the uranium and migrated. Eventually it will decay into ^{235}U at its new site. Hence the uranium here will accumulate excess ^{235}U, and its isotopic composition will show an enrichment of ^{235}U. As stated above, however, compared to natural uranium the isotopic compositions of uranium quarried from the Oklo mine areall deficient in ^{235}U, and no uranium with an excessconcentration of ^{235}U has been found. This demonstrates that the ^{239}Pu formed in the Oklo natural nuclear reactor did not separate from the ^{235}U that produced it, but acted together with it. It is unlikely that U and Pu behave in exactly the same manner when element movement occurs as the result of rock weathering and other geological disturbances, and so most likely this indicates that the movement of ^{239}Pu itself was negligible. If this is the case, it will mean that ^{235}U did not move significantly from its original location.

A similar conclusion can be reached for the rare earth elements formed in the natural nuclear reactor. The fission products of ^{235}U also include Nd, Sm, and other rare-earth elements. In general, when nuclei undergo nuclear fission they produce many fission products, but the type of nuclei produced and their proportion depend on the kind of nucleus involved in the fission. This is known as the fission yield. Take the three rare earth elements of Nd, Sm, and Gd, for instance. If we consider the isotope compositions of ^{143}Nd + ^{144}Nd, ^{147}Sm + ^{148}Sm, and ^{147}Gd + ^{158}Gd, the proportion formed through the nuclear fission of ^{235}U is 1 : 0.203 : 0.00142. The value in rocks collected from the core of the Oklo natural nuclear reactor is 1 : 0.204 : 0.0014 on the average, and when experimental error is taken into account, this can be regarded as identical to the fission yield of ^{235}U. This probably indicates that Nd, Sm, Gd, and other such rare elements did not move from the site of nuclear fission. If weathering had caused the Nd, Sm, and Gd to move after fission, each element would behave in a slightly different manner reflecting their differences, and so their isotopic composition would differ from the fission yield.

But not all fission products remained near the core of the natural nuclear reactor. As an extreme case, it has been confirmed that Xe, Kr and other rare gases that are a fission product of ^{235}U have escaped almost completely from the rocks near the core. It has also been concluded that a considerable proportion of elements such as Rb, Sr, Cd, and Ba, which from a geochemical viewpoint escape relatively easily from rocks during weathering, have also been lost from the core part. However, no exact estimate has been made yet of how far these elements moved and how long this movement took. The results of research so far suggest that the uranium, plutonium, and other elements forming the core of the nuclear reactor were hardly dispersed at all around the reactor, and have been preserved as is for 2000 million years, but that a considerable number of other elements have moved away from the core. Considering that 2000 million years have elapsed since the Oklo natural nuclear reactor first went into "operation", and that its core is comparatively close to the surface of the earth and hence prone to the effects of weathering, the Oklo nuclear reactor corresponds to quite a severe case of discharge of radioactive elements.

The Oklo nuclear reactor offers an incomparable case study for forecasting the behavior of radioactive waste. It is hoped that more detailed research will be carried out in the future, including quantitative estimates of the movement of various elements and studies of the influence of radiation on the rocks surrounding the core, particularly its effect on the weathering of rocks. Surveys of the geological stability of proposed waste disposal sites and research into rock stability vis-à-vis radioactive nuclides are also essential measures for dealing with radioactive waste. Investigations on a geohistorical time scale will be necessary in order to confirm the geological stability of these sites. After more than 5 years of investigation, the Working Group on the Disposal of High-Level Nuclear Wastes established within ICSU (International Council of Scientific Unions) in 1978 emphasized the importance of geological studies of proposed waste disposal sites in addition to material science research into the receptacles for accommodating radioactive waste – past research has been largely confined to this aspect. Confirmation of the geological stability of these sites by using the geohistorical time scale will be the most vital of these geological studies. Geohistorical research is expected to make a major contribution to a solution of such problems in the future.

5.3 Epilog

Most geohistorical phenomena materialize over millions or hundreds of millions of years. Owing to this enormous time span, it is obviously impossible to reproduce and observe these phenomena in the laboratory. Not only does the time involved far exceed man's experience, but time is also an essential part of these geohistorical phenomena. It would be virtually meaningless to attempt to grasp these phenomena by looking only at their static features and not taking time into account. Geohistorical phenomena can only be understood by resolving the dynamic aspects of the earth as it has changed over time. Compare this situation with the common methods used in physical and chemical research. In these fields complex phenomena are broken down into several simple elementary processes and analyzed, and then an attempt is made to comprehend the whole as the aggregate of these processes. This method conventionally used in physics and chemistry, however, is not necessarily useful in elucidating geohistorical phenomena. Here we will use the example of the origin of petroleum to explain the essential importance of this vast time.

Though many questions remain about the origin of oil, the generally accepted scenario is that organic matter buried underground underwent organic metamorphism under appropriate temperature and pressure conditions, and thus formed the hydrocarbon molecules constituting petroleum. The hydrocarbon molecules produced at this time and the quality of the petroleum itself vary depending on differences in the extent of organic metamorphism. A. Hood and his colleagues introduced the LOM (Level of Organic Metamorphism) index as a parameter for organic metamorphism. On the LOM index 0 corresponds to the lowest degree of organic metamorphism and 20 to the highest value. Similar to inorganic metamorphism, the higher the temperature, the more rapidly the organic metamorphism reactions proceed. Consequently, the higher the temperature, the less time it will take to obtain the same metamorphic effect. Hood and his colleagues inferred the actual conditions under which petroleum is produced, i.e., the temperature at which organic metamorphism occurs and its duration, and compared

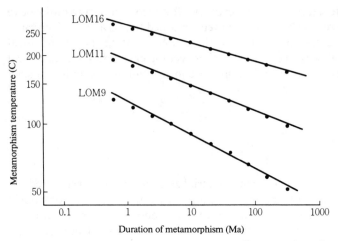

Fig. 5.4. Temperature of organic metamorphism for generation of petroleum versus time duration of the metamorphism. *LOM* stands for Level of Organic Metamorphism, which gives a measure of the quality of petroleum. The slope of the lines is proportional to the activation energy characteristic to the organic metamorphism. (After Hood et al., 1975)

these data with the LOM of that petroleum to produce Fig. 5.4. Here the vertical axis denotes the temperature (scaled as 1/T) at which organic metamorphism occurs, and the horizontal axis denotes the duration. As Fig. 5.4 reveals, for a certain LOM value the time duration and the reciprocals of the temperature show a neat linear correlation sloping down toward to the right. Note that oil of a certain quality, i.e., oil corresponding to a particular LOM value, is produced only through the special combination of temperature and time shown in Fig. 5.4.

The neat linear relations in Fig. 5.4 indicate that organic metamorphism follows the Arrhenius relation

(Reaction velocity) $\alpha \exp(-E/RT)$. (5.11)

Here E, R, and T indicate respectively the activation energy, gas constant, and temperature of the reaction. Since the reciprocal of the reaction velocity on the lefthand side can be regarded as the duration of the reaction, if this time is denoted as γ, then from Eq. (5.11)

$\ln \gamma = A - E/RT$ (A: constant)

This shows that the logarithm of the reaction time (γ) has a linear relation with the reciprocal of the temperature (T), and the activation energy can be deduced from the slope of this line.

Figure 5.4 shows that in organic metamorphism time has the same effect as temperature. Initially this may raise hopes that if the effect of time can be grasped through temperature, that is, by raising the temperature and increasing the reaction velocity, it may be possible to produce in the laboratory petroleum that is identical to that produced over geological time. However, this is not the case. Increasing the reaction velocity to match the laboratory time scale would require a high temperature, as can be deduced from the extension of the straight line in Fig. 5.4. At this temperature, however, the hydrocarbon molecules would decompose.

Let us discuss this situation in somewhat general terms using an Arrhenius plot. Suppose that process A (e.g., the organic metamorphism that produces petroleum molecules) is specified by the activation energy (E_A), and that it follows the Arrhenius equation. We then suppose that process B (e.g., the decomposition of petroleum molecules) also follows the Arrhenius equation, and that it has activation energy (E_B). Figure 5.5

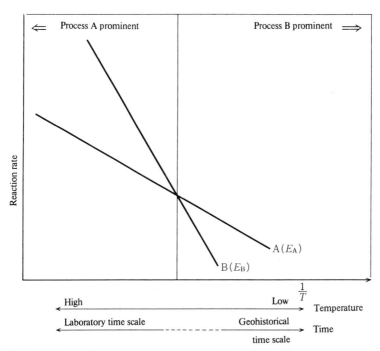

Fig. 5.5. Figure illustrates the importance of time in geohistorical phenomena. An attempt to promote a process *A* (e.g. generation of petroleum) by raising temperature must promote a process *B* (e.g. thermal decomposition of petroleum), which will dominate the process A at higher temperature. Hence, the process A is visualized only at relatively low temperature, which then necessarily require a geohistorical time span, that is to say, an enormous time span is an essential factor in many geohistorical phenomena (e.g. generation of petroleum). E_A and E_B are activation energies which characterize the process A and B.

showsthe general outline of this situation. At relatively low temperatures (T) process A (e.g., the formation of petroleum molecules) is more rapid than process B, but when the temperature is increased so as to correspond to laboratory time (τ), process B (the decomposition of petroleum) excels, and process A (the production of petroleum) becomes impossible. Time may be regarded as equivalent to the temperature for each single process, and the reaction time can be reduced by raising the temperature. Hence in natural phenomena that materialize as the combination of numerous processes, the time-temperature combinations that will fit all of these processes in order for the phenomena to materialize are extremely limited. This example demonstrates clearly that phenomena produced by nature over many long years cannot necessarily be reproduced easily in the laboratory.

Here we have discussed from a kinetic viewpoint the significance of the long geological time that characterizes geohistorical phenomena. Let us next consider this significance from the aspect of the statistical nature that generally governs geological phenomena. As stated in Chapter 4, at present we have no clear understanding of the driving force behind the movement of the plates. From the results of paleomagnetic research and the magnetic anomaly patterns in the oceanic crust, however, we can conclude fairly certainly that plate movements have occurred on a time scale of millions or tens of millions of years. Many different factors are involved over an enormous time span, so systematic movement – the plates moving away at right angles from the mid-ocean ridge toward subduction – cannot be recognized until the plate movements are viewed as an average on this long time scale. Perhaps plate movements and continental drift are only conceivable when observed in line with this enormous geohistorical time scale.

Let us take a familiar example. Suppose that we fill a basin with water and drop some red ink into the water. As time passes, the ink will gradually spread throughout the water. It would be out of the question for it to reverse direction and converge again into the original drop of ink. If the movements of each individual ink molecule are observed over a very short period, however, some of them would even be observed moving back to the original position. As the average of the interaction of countless numbers of water molecules, however, the ink molecules spread out uniquely from the point where the ink was dropped, following the law of diffusion. Perhaps plate movements should be viewed in the same light. Observed in accordance with our empirical time measure, which is extremely short compared to the geohistorical time scale, the plates seem to be moving at complete random. In recent years attempts have been made to prove the existence of plate movements by using laser beams and radio wave interference, but no matter how precise these observations may be, they may be of little use if plate movements are essentially geohistorical phenomena that cannot be apprehended as a momentary image, and if their real image only appears as an average of many long years.

The significance of the geological time scale that is the most fundamental characteristic of geohistorical phenomena can hardly be overemphasized. Applying conventional approaches that have been enormously successful in physical and chemical research may not be very rewarding when dealing with geohistorical phenomena. A different approach must be sought to understand geohistorical phenomena, few significant results can be expected. An original method is necessary in order to understand these phenomena. This is the method of seeking in nature "fossil" records of geohistorical phenomena, and using these to throw light on these phenomena. As examples of this, we have seen how the remanent magnetization in rocks and the magnetic anomaly lineation on the sea floor contain vital information on continental drift and plate movements. We also learnt how radiogenic isotopes provide invaluable information as a tracer of material movements in the earth's interior that occur on a global scale and in accordance with the geological time measure. Geohistorical research occupies a unique position in the earth sciences in two respects – the fact that its subject is phenomena measured on a long time scale corresponding to the whole history of the earth, and the fact that the methods used seek "fossils" of these phenomena in nature. Owing to its "historical" nature, geohistorical research provides us with a very useful lead to forecasting the future of the earth. Geohistory is still a fledgling discipline, but it seems to hint at its future as a vital field in earth science.

Bibliography

Anders E, Ebihara M (1982) Solar-system abundances of the elements. Geochim Cosmochim Acta 46:2363–2380
Armstrong RL (1981) Radiogenic isotopes: the case for crustal recycling on a near-steady no-continental-growth earth. In: Moorbath S, Windley BF (eds) The origin and evolution of the Earth's continental crust. The Roy Soc Lond, pp 259–287
Black DC, Pepin RO (1969) Trapped neon in meteorites-II. Earth Planet Sci Lett 6:395–405
Braginsky SI (1964) Kinematic models of earth's hydromagnetic dynamo. Geomag Aeron 4:732
Broecker WS, Oversby VM (1971) Chemical equilibria in the earth. McGraw-Hill Book, New York, p 318
Brown H (1952) Rare gases and the formation of the earth's atmosphere. In: Kuiper GP (ed) The atmospheres of the earth and planets, 2nd ed. Univ Chicago Press, pp 258–266
Brown L, Klein J, Middleton M, Sacks, S, Tera F (1982) ^{10}Be in island arc volcanoes and implications for subduction. Nature 299:718–720
Clayton RN, Onuma N, Mayeda T (1976) A classification of meteorites based on oxygen isotopes. Earth Planet Sci Lett 30:10–18
Compston W, Pidgeon RT (1986) The age of (a tiny part of) the Australian continent. Nature 317:559–560
Creer KM (1967) A synthesis of world-wide paleomagnetic data. In: Runcorn SK (ed) Mantles of the earth and terrestrial planets. Interscience, Lond, pp 351–382
Davidson JP (1983) Lesser Antilles isotopic evidence of the role of subducted sediments in island arc magma genesis. Nature 306:253–256
DePaolo DJ (1981) Nd isotopic studies: some new perspectives on Earth structure and evolution. EOS Rev Esp Entomol 62:137–140
DePaolo DJ (1983) The mean life of continents: estimate of continent recycling rates from Nd and Hf isotopic data and implications for mantle structure. Geophys Res Lett 10:705–708
Ebihara M (1984) Terrestrial and solar abundances of elements (in Japanese) Chishitsu News. Geol Surv Jpn Rep, Sept. 8–19
Fanale FP (1971) A case of catastrophic early degassing of the earth. Chem Geol 8:79–105
Fyfe WF, Babuska V, Price NJ, Schmid E, Tsang CE, Uyeda S, Velde B (1984) The geology of nuclear waste disposal. Nature 310:537–540
Gamov G (1946) Expanding universe and the origin of elements. Phys Rev 70:572–573
Ganapathy R (1980) A major meteorite impact on the earth 65 million years ago: Evidence from the Cretaceous-Tertiary boundary clay. Science 209:921–923
Ganapathy R, Anders E (1974) Bulk compositions of the moon and Earth, estimated from meteorites. Geochim Cosmochim Acta 5:2, 1181–1206
Gast PW (1960) Limitations on the composition of the upper mantle. J Geophys Res 65:1287–1297

Gauthier-Lafaye F (1978) Suivi geologique de l'exploitation des reacteurs naturels d'OKLO IAEA-TC-119/3 B. UNESCO Tech Pap Mar Sci, Paris

Goldreich P, Ward WR (1973) The formation of planetesimals. Astrophys J 183:1051–1061

Grossman L (1972) Condensation in the primitive solar nebula. Geochim Cosmochim Acta 36:597–619

Hamano Y, Ozima M (1978) Earth-atmosphere evolution model based on Ar isotopic data. In: Alexander EC jr., Ozima M (eds) Terrestrial rare gases. Cent Acad Japan, Tokyo, pp 155–171

Hayashi C (1972) Origin of the solar system. Proc. 5th Lunar Planetary Symposium, 13–18. Inst Space Aeronautical Sci, Univ Tokyo

Hayashi C, Nakazawa K, Nakagawa Y (1985) Formation of the solar system. In: Black DC, Matthews MS (eds) Protostars and Planets II: Univ Arizona Press. Tucson, Arizona, pp 1100–1151

Hohenberg C, Podosek FA, Reynolds JH (1967) Xenon-iodine dating: sharp isochronism in chondrites. Science 156:202–206

Holland HD, Lazor B, McCaffrey M (1986) Evolution of the atmosphere and oceans. Nature 320:27–33

Holmes A (1946) An estimate of the age of the earth. Nature 157:680–684

Hood A, Gutjahr CCM, Heacock RL (1975) Organic metamorphism and the generation of petroleum. Am Assoc Pet Geol Bull 59:986–996

Houtermans FG (1953) Determination of the age of the earth from the isotopic composition of meteorite lead. Nuvo Cimento 10:1623–1633

Itoh E, Harris D, Anderson AT jr. (1983) Alteration of oceanic crust and geologic cycling of chlorine and water. Geochim Cosmochim Acta 47:1613–1624

Jeffrey PM, Reynolds JH (1961) Origin of excess Xe^{129} in stone meteorites. J Geophys Res 66:3582–3583

Jones DL, Cox A, Coney P, Deck M (1982) The growth of Western North America. Sci Am 247, 5:70–84

Kent DV (1981) Asteroid extinction hypothesis. Science 2111:648

Kuroda PK (1956) On the nuclear physical stability of the uranium minerals. J Chem Phys 25:781–782

Larson RL (1981) Geological evolution of the Nauru basin, and regional implications. In: Initial Report of Deep Sea Drilling Project. Washington DC (US Gov Printing Office), 61:841–862

Lewis JS, Prinn RG (1984) Planets and their atmospheres – origin and evolution –. Academic Press, London, p 470

Long LE (1964) Rb–Sr chronology of the Carn chuinneag intrusion. Rosshire, Scotland. J Geophys Res 69:1589–1597

McElhinny MW, Senanayake WE (1982) Variations in the geomagnetic dipole 1: The past 50,000 years. J Geomag Geoelectr 34:39–51

Merrill RT, McElhinny MW (1983) The Earth's magnetic field. Academic Press, London, p 401

Miller SL (1953) Production of amino acids under possible primitive earth conditions. Science 117:528–529

Moorbath S (1977) The oldest rocks and the growth of continents. Sci Am 236:92–104

Morgan J (1972) Convection plumes and plate motions. Am Assoc Pet Geol Bull 56:203–213

Néel L (1949) Theorie du trainage magnétique des ferromagnétiques aux grains fins avec applications aux terres cuites. Ann Geophys 5:99–136

Niemeyer S (1978) I–Xe dating of silicate and troilite from IAB iron meteorites. Geochim Cosmochim Acta 43:834–860

Officer CB, Drake CL (1985) Terminal Cretaceous environmental events. Science 227:1161–1167
Oversby VM, Ringwood AE (1971) Time of formation of the earth's core. Nature 234:463–465
Ozima M, Podosek FA (1983) Noble gas geochemistry. Cambridge Univ Press, Cambridge, p 367
Ozima M, Zashu S (1983) Primitive He in diamonds. Science 219:1067–1068
Patterson C (1956) Age of meteorites and the earth. Geochim Cosmochim Acta 10:230–237
Peltier WR (1980) Mantle convection and viscosity. In: Dziewonski AM, Boschi E (eds) Physics of the earth's interior. Elsevier, New York, pp 362–431
Penzias AA, Wilson RW (1965) A measurement of excess anntenna temperature. Astrophys J 142:419–421
Poldervaart A (1955) Chemistry of the earth's crust. Geol Soc Am Spec Pap 62:119–144
Reynolds, JH (1960) Isotopic composition of primordial xenon. Phys Rev Lett 4:351–354
Richardson SH, Gurney JJ, Erlank AJ, Harris JW (1984) Origin of diamonds in old enriched mantle. Nature 310:198–202
Ross JE, Aller LH (1976) Chemical composition of the sun. Science 191:1223–1229
Rubey WW (1951) Geological history of sea water: an attempt to state the problem. Geol Soc Am Bull 62:1111–1147
Safronov V (1969) Evolution of the protoplanetary cloud and formation of the earth and planets. NASA TTF-677, US Department of Commerce
Sasaki S, Nakazawa K (1986) Metal-silicate fractionation in the growing earth. J Geophys Res 91: 9231–9238
Shidlowski M (1983) Evolution of photoautotrophy and early atmospheric oxygen level. Precambrian Res 20:319–335
Schilling JG (1973) Iceland mantle plume: geochemical evidence along Reykjanes Ridge. Nature 242:565–571
Schneider SH, Mass C (1975) Volcanic dust, sun spots, and temperature trends. Science 190:741–746
Shoemaker EM (1983) Asteroid and comet bombardment of the earth. Ann Rev Earth Planet Sci 11:461–494
Steuber AM, Ikramuddin M (1974) Rubidium, strontium and the isotopic composition of strontium in ultramafic nodule minerals and host basalts. Geochim Cosmochim Acta 38:207–216
Takaoka N (1982) Noble gases in meteorites (in Japanese). The Astronomical Herald 75:199–204
Tauxe L, Tucker P, Petersen NP, LaBreque JL (1984) Magnetostratigraphy of Leg 73 sediments. In: Initial Reports of the Deep Sea Drilling Project. Washington DC (US Gov Printing Office), 73:609–621
Taylor SR (1977) Island-arc models and the composition of the continental crust. In: Ewing Series Vol 1. Am Geophys Union, Washington DC, pp 325–335
Théllier E, Théllier O (1959) Sur l'intensité du champ magnétique terrestre dans le passé historique et géologique. Ann Geophys 15:285–376
Vine FJ, Matthews DH (1963) Magnetic anomalies over oceanic ridges. Nature 199:947–949
Vollmer R (1977) Rb isotope chronology of core formation. Nature 234:144–147
Wetherill G (1975) Radiometric chronology of the early solar system. Ann Rev Nucl Sci 25:283–328
Wilson JT (1963) A possible origin of Hawaiian islands. Can J Phys 41:863–870

Subject Index

accretion of the earth 64
 energy 58
 heterogeneous 67
 homogeneous 64, 66–67
accretionary prism 119, 121
age of degassing 90
age of meteorite 30–34
^{26}Al 14
Allende meteorite 17
Antarctica 40
^{40}Ar/^{39}Ar dating method 125
^{40}Ar/^{36}Ar ratio 89
 in the mantle 90, 91
Arrhenius equation 153
atmosphere 85, 89
 blanket effect 65
 primary 27, 86–88
 secondary 27, 86–87
aubrite 37

banded iron formations 93
^{10}Be 121
big bang 6, 7
binary stars 24
biosphere 85
bolide 136

C (carbon) 92, 93
 in Isua metamorphic rock 93
C1 chondrite 20, 136
carbonaceous chondrite 16, 17, 18, 38
catastrophic degassing 90
CH_4 92
chondrites 74
 enstatite 37
 equilibrium 32
Co 140
CO_2 134
CO_2 well gas 96
comparative planetology 43
condensation of elements 22, 40–41

condensation theory 21
 nonequilibrium 24
continental drift 1, 115, 117, 118, 154
 driving force of 118
continuous degassing 90
continuous degassing model 86
core 44, 62, 109, 110
 differentiation 44, 58
 formation 29, 65
 inner core separation 58
 Mars 45
core-mantle separation 58
 age of 70
Cretaceous-Tertiary boundary (also see K–T boundary) 135
crust 62, 63, 77
 composition 30, 63
 differentiation of 94
 formation 79–80
 oldest 131
 Pb in 71
crustal formation 76
 continuous 72
cryogenic magnetometer 101

D (deuterium) 104
decay constant 123
declination 104
deep sea sediment 52
 magnetic minerals in 101
deepest bore hole 59
degassing 1, 41
 age of 90
 atmosphere 89
 catastrophic 90
 continuous 90
 rare gas 87
depositional remanent magnetization (DRM) 101, 104
diamond 81, 82
 ^{3}He in 83
 inclusions in 81

dip 104
disposal of nuclear waste 143
distribution coefficient 67, 69
  of Ni 67
dynamo theory 2, 109

Earth 41, 45
  accretion energy 55, 58
  composition 20, 30, 59, 89
  energy source in 58, 64
  formation 28
  layered structure 28
  magnetic field 2
  primitive (or primeval) 70, 71
  rotation axis of 105–106
earth's magnetic field 2
  intensity 108
  polarity 102, 103
  reversal of 102, 104
  temporal variation 108
El Chichon volcano 141
electron capture nuclear
  disintegration 89
elemental fractionation 76
enstatite chondrite 37
$\varepsilon_{Nd} - \varepsilon_{Sr}$ diagram 78, 79
equilibrium chondrites 32
equilibrium constant 21
escape velocity 40
exotic terrains 117
explosion of a supernova 25
extinct nuclide 35
extraterrestrial material 136

Fe 62
  formation of 7
Fig Tree shale 93
fissiogenic Nd 145
formation interval (or period) 34, 36, 37
fossil 2, 99
  geomagnetic field 100
fragmentation of the dust layer 26

geocentric axial dipole field 107
geocentric dipole field 104, 106
geohistorical phenomena 155
geomagnetic dipole 43
geomagnetic dipole axis 105
geomagnetic field
  (see Earth's magnetic field)
geomagnetic polarity time scale 112
geomagnetic pole 105

geotherms 82
granite 72
graphite 82
graphite-diamond phase diagram 82

H (hydrogen) 7
He (helium) 7
  isotopic composition 94
  primitive 83
  primordial 95
^{3}He 83
  in diamonds 83
^{3}He/^{4}He ratio 83
  primordial 83–84
heat flow 60
heterogeneity of solar material 17
history 3
H_2O 93
hot spot 94

^{129}I 9, 12, 18, 34, 35, 37, 47, 51, 95
impact degassing 90
impact of bolide 136, 137
inclination 104
International Phase of Ocean Drilling
  (IPOD) 113
interstellar clouds 21
interstellar gas 8
Ir 134, 135, 139, 140
isochron method 125
  Rb–Sr 8, 33, 127
isochron plot 32
isotopic anomaly 17
isotopic compositions 14
  planetary atmospheres 41
  solar materials 15
isotopic heterogeneity 37
Isua metamorphic rock 46, 93, 110
I–Xe formation age 37, 50

K (potassium) 56, 123
  in the mantle 91
K–Ar dating method 123, 124, 126, 128
K–T boundary (also see Cretaceous-
  Tertialy boundary) 134, 136

lead ores 52
LOM (Level of Organic
  Metamorphism) 151, 152
Lu–Hf systematics 128
lunar cataclysm 38

magma ocean 28
magnetic anomaly 111
   lineation 111
   pattern 114
magnetic fields of planets 43, 45
magnetic poles 105
magneto-hydrodynamics 44
mantle 62
   composition 63
   K-content 91
   lower 63
   Ni in 67
   primitive (or primordial) 76, 80, 94–95
   undifferentiated 80, 94
   upper 63
mantle array 79
mantle convection 80, 83, 118
mantle-crust system 72, 77
mantle evolution 76
marine regression 122
Mars 39, 44, 45
mass extinction 134, 140
mass fractionation 16
Mercury 39, 41, 45, 59
metamorphism 126
meteorites 5, 16
   age of 17, 37
   collision with earth 138
   elemental abundance 18
   formation of 10, 14
   Ne in 17
   of lunar origin 38
   oxygen isotopes 15–16
   parent bodies of 29, 59
   primitive 65
   undifferentiated 20
meteorite analogy 20, 30, 46–47, 59, 60, 78, 84
meteorite crater in Arizona 139
meteorite shower 38
microtektite 136
Mid-Atlantic Ridge 111
mid-ocean ridge 119
Miller's experiment 92
mineral age 32, 127
Mn nodules 71
Moon 38–39
   composition 61
MORB 70, 79, 94, 95, 120, 129
   Pb in 52

N (nitrogen) 91
natural nuclear reactor 144
Nd 73
Nd isotopic ratio 75, 145
   evolution of 74
Ne 17
Ne–E 17
nebulae 14
nebular gas 14
neutron capture reaction 8
NH_3 92
Ni 66
   in the mantle 67
noble metal 135
non-volatile elements 6, 19
nuclear fission 144, 147
nuclear winter 141
nucleosynthesis 9, 10, 11, 12, 13, 18, 34, 35, 47
   continuous 10–12
   p-process 8
   r-process 8
   s-process 8
   sudden 10–13

O (oxygen) 6, 16–17, 91
ocean floor spreading theory 111
ocean trench 119
Oklo natural reactor 143, 146, 148
oldest crustal rocks 93, 110, 129
oldest crust 131
ordinary chondrite 37
organic metamorphism 151, 152
origin of the atmosphere 1
origin of life 92
origin of the solar magnetic field 109
oxygen isotopic ratio 39
   in meteorites 15–16

paleo-dipole moment 108
paleomagnetic methods 99
paleomagnetism 3
paleosols 93
partial melting 76
Pb 68
   in MORB 70
Pb–Pb method 34
Pb isotopic ratio 51, 52
   in the earth 68
petroleum 151, 152
p-process 8
planetary gases 95
planetary Xe 95

planetesimal 26, 27, 55, 64, 65, 87
  collision of 88
plate movement 154
plate tectonics 43, 110, 114
plutonium 149
polar wandering curve 106, 115, 116
pre-solar materials 18
primary atmosphere 27, 86, 88
  composition of 91
primary magnetization 102
primitive meteorites 65
primitive (or primeval) earth 70, 71
primitive He 83
primitive solar nebula 21
  pressure in 40
primitive solar system material 18
primitive sun 25
primordial He 95
primordial lead 69
proto-planets 26
^{244}Pu 9, 12, 34–35, 95

radioactive decay 58, 123
  discovery of 122
radioactive energy 29, 56
radiometric dating method 123
rare gases (or noble gases) 18, 41, 85
Rayleigh number 118
Rayleigh-Taylor instability 66
Rb 31, 56, 60, 77, 125
Rb–Sr dating method 125–127
  isochron plot 81, 127
  mineral age 127
  of metamorphic rocks 33
  whole rock age 127
recycling hypothesis 72
reducing atmosphere 92
reversal of geomagnetic field 102, 104
r-process 8

s-process 8
secondary atmosphere 27, 86, 87
secondary remanent
  magnetization 102
sharp isochronism 36
shergottite 39
siderophile elements 66
silicate inclusions in iron
  meteorite 37
Sm 75
Sm–Nd dating method 128–129
SMOW 15
solar abundance 21, 22

solar constant 57
solar energy 57, 58
solar material 16
solar nebula 16, 17, 20, 25, 28, 29, 39, 46, 59, 85
  chemical composition 60
  dissipation of 26
  early 18
  elemental abundance 20
  P–T distribution in 24, 40
  temperature 23, 24, 25
solar nebular gas 88
solar photosphere 19
solar radiation 57
solar system 14, 25
  elemental abundance in 8, 18, 20
  formation of 14, 18
solar wind 87
Sr 31, 60, 77
^{87}Sr/^{86}Sr ratio 130
  in solar nebula 33
  initial 31–32
St. Helens eruption 141
stellar explosion 10
step-wise degassing 49, 50
subduction 72, 119, 120, 121
sun 14, 18
  corona 14, 18
  elemental abundance in 18
  primitive 25
  radiative energy from 57
supernova 14

T Tauri stage 87, 88
Tambora volcano 141
terrestrial planets 39
  formation of 26–27
Th 68, 94
Thellier's method 107
thermo-remanent magnetization
  (TRM) 100
  artificial 107
  stability of 100
titanomagnetite 99
Toba volcano 140
trace elements 76
troilite 69
  in iron meteorites 37
Tunguska fireball 138–140

U (Uranium) 9, 12, 13, 56, 68, 94, 143
  anomalous isotopic composition
    of 145

Subject Index

ulvospinel 99
undifferentiated meteorites 20
undifferentiated mantle 80, 94
uniformity of the isotopic composition 59
upper mantle 63
UR (uniform reservoir) 78

Venus 39, 45
VGP 105, 115
Vine-Matthews hypothesis 113
virtual geomagnetic poles (see VGP)
viscous remanent magnetization (or VRM) 102

volatile component 41, 85, 86
　in the primitive earth 92
volatile elements 19, 60

whole rock age 127

Xe 48, 95
　primordial 51
^{129}Xe 47–51
xenolith 59, 80

Yamato base 39

zircon 110, 131